KENKEN
E L I T E

Why Settle For Expert?

100 UNBELIEVABLY CHALLENGING PUZZLES THAT MAKE YOU SMARTER

CREATED BY
TETSUYA MIYAMOTO

To access free, unlimited puzzles of all sizes and difficulty levels, visit **www.kenkenpuzzle.com**

Download the **FREE** KenKen Classic app for iOS, Android or Kindle Fire in the App Store, Google Play Store, or Amazon App Store

KenKen Elite: Why Settle For Expert?
Puzzle contents copyright ©2018 KenKen Puzzle, LLC. All rights reserved.
KenKen is a registered trademark of KenKen Puzzle, LLC. All rights reserved.
www.kenkenpuzzle.com

Project editors: Tyler Kennedy, Jerry March
Designer: Susannah Fears

ISBN-13: 978-1-945542-08-4
ISBN-10: 1-945542-08-X
First Edition: February 2018

INTRODUCTION

ELITE

Noun - /e'lēt, ā'lēt/ - *a select part of a group that is superior to the rest in terms of ability*

If, by definition, you believe that your KenKen skills align with this group, then you should ask yourself - *Why Settle For Expert?* "Kenerated" by our crack production team with only true KenKen masters in mind, each puzzle has been specially formulated to maximize both the difficulty and the reward. If you can solve them, you can - and should - consider yourself one of the very few KenKen Elite!

Developed in 2004 by renowned educator Tetsuya Miyamoto, KenKen has quickly become the world's most addictive and fastest growing puzzle since Sudoku. Miyamoto-sensei's goal was to improve not only his students' math and logic skills, but also to encourage independent thinking by emphasizing creativity, concentration, and perseverance. Little did he imagine that his simple yet sophisticated (and unbelievably fun!) puzzle would transcend the classroom and become a worldwide sensation. Now published daily in over 150 newspapers and magazines worldwide including The New York Times, The Times (UK), Spiegel Online (Germany), El Pais (Spain), The Globe and Mail (Canada), the Los Angeles Times, and many others, KenKen is truly a global phenomenon. Over 50 million puzzles are played online each year, with hundreds of millions more solved in print and on mobile devices.

No doubt, as an experienced KenKen solver, you are adding to these totals. And while online play is great in the home or office and nothing beats conquering KenKen in the newspaper during your morning commute, KenKen books are perfect almost anywhere...planes and trains, bedrooms and bathrooms. So we are pleased to present you with the book you now hold in your hand: **KenKen Elite:** *Why Settle For Expert?* ... the first in the KenKen Puzzle Company's official series of "harder than hard," "more expert than expert" puzzles.

These intricate and intriguing KenKens are designed to unlock the most challenging experience for any master KenKen solver. Complete them all and you will stand beside the most accomplished puzzle minds from all corners of the world.

While it's true that KenKen was originally invented to educate and exercise the mind, its main goal is to entertain. So sit back, grab a pencil, and start solving. Enjoy!

How To Solve

At its core, KenKen is a simple yet rich logic puzzle with easy-to-understand rules:

◊ Fill each square in the grid with a single number. In a 9x9 grid, use the numbers 1 through 9.

◊ Do not repeat a number in any row or column. For example, in a 9x9 grid, each row and each column should be filled in with the numbers 1 – 9 with no duplication.

◊ Each heavily outlined set of squares is called a "Cage." The numbers in each Cage must combine – in any order – to produce the Target Number indicated in the top corner of the Cage by using the math operation next to the Target Number. For example, if the Target Number shows "18x" and there are 3 squares in the Cage, the numbers in that Cage must total 18 in any order by using multiplication (9x2x1, 6x3x1, 3x2x3, etc.).

◊ A number may be repeated within a Cage as long as it is not in the same row or column. You can do this when a Cage has an "L" shape or spans several rows and/or columns.

Hint: To start, fill in the single-square Cages first. Since no calculation is required and the answer is "given" to you, we call these "Freebies."

And that's it!

Solving a KenKen puzzle involves pure logic and mathematics. No guesswork will ever be needed, and each puzzle has only one solution. So sharpen your pencil, put on your thinking cap, and find out why Will Shortz, NPR Puzzle Master and The New York Times Puzzle Editor calls KenKen "The most addictive puzzle since Sudoku!"

Visit www.kenkenpuzzle.com any time to play unlimited puzzles of all sizes and difficulty levels, absolutely free of charge.

+ − × ÷

8×	11+		17+	24×	3−		4−	
		3÷			1−	243×		2÷
2−			18×					
15+	4−	7+			3÷	96×		
		8×		6−			108×	
	2÷	22+			135×			
18×			12+	3	12+	11+		
	22+						42×	
3÷			2÷		1−		9+	

SOLVING
TIME

+ − × ÷

3÷		1−	3−	3−	4−	196×		5−
13+	15+					3−		
		4−		21×			405×	8×
		40×	2÷		2−			
42×				19+		5+		3÷
18+		12+				21+	10+	
		12×	8+	48×				23+
18+				1−	9+			
	2÷		9					

SOLVING
TIME

+ − × ÷

15×		9+	168×		4÷		10+	
3÷			21+		120×			13+
5	168×				18+	84×		
4÷		4−					7+	
	7−	1−	8+				5−	2−
16+			1−		10+			
	16+		4÷		1−	15×	35×	2÷
24×		1−	2−	2÷				
					6−		3−	

SOLVING
TIME

+ − × ÷

2÷	18×		3÷		2−		13+	
	2−		245×	90×		144×		7−
13+					3			
40×	3÷	1−		5−		2−		20+
		10+		7−		13+		
	42×	22+		11+				
			30×		252×		17+	12+
26+		48×			22+			
						1		

SOLVING TIME

5

+ − × ÷

180×			14+	4÷	13+		16+	
30×	84×					3−		36×
		2÷	13+				1−	
			378×		5−			3÷
16+		5−			13+		30×	
	216×	14×		20×		6×		14+
			160×	1134×				
17+	11+					3−		5−
			6+		14+			

SOLVING
TIME

6

$+ - \times \div$

6+		3÷	1944×			168×		
144×				4÷		7−	2−	
	3÷		14×	18+			180×	
11+		20×				15+		5−
4÷	20+			13+	7			
			4÷		1−		2÷	7−
16+				9	13+			
2−	7−		3÷		13+		1−	5−
	3−		2−		7+			

SOLVING
TIME

+ − × ÷

16×			2÷		19+	140×	3−
108×	13+		1−	5−			8−
		4÷				3÷	24×
	7		4−		24+		
30+		84×		1−			2−
	48×			15+	5−		24×
						21+	
3÷		36×		189×			240×
17+					2÷		

SOLVING
TIME

+ − × ÷

1−		10+		5−	4−	64×		
5−		28×				3÷	10+	
21+		8	3−		168×			1−
80×		20+				4	6×	
	8×		4−			18+		6−
	17+		48×	24×	1		2−	
		72×						2÷
13+			2−	2÷		4−		
				108×			7−	

SOLVING
TIME

9

$+ \; - \; \times \; \div$

3	40×		10+		17+			140×
10+		4−		4−		2−		
4−			180×				7+	
56×		20+	6−	2	2−			
11+				21+	2÷	10+	4−	
	84×						12+	
		2÷	160×	19+	10×	4−		
2−	10+					1−		24×
						11+		

SOLVING TIME

+ − × ÷

2÷	30×			22+		9	5−	7−
	1−	7+			2−			
15+		28×		1−	1−	60×		13+
	21+		24×				8	
				17+	15+	4÷	4−	
3−							63×	
13+		180×	8+			7+		3
168×			7−	3−	8+		2−	
					15+		12×	

SOLVING
TIME

+ − × ÷

7−		48×		5	24×		26+	
80×			3÷		19+			
	2−		35×	6−	2÷		20×	3−
15×	1−	2−						
		22+	12×			360×	2÷	
5−					1−			
2−	15×		3−				48×	
		7−		20×		1−		
5−		16+			3÷		2−	

SOLVING
TIME

+ − × ÷

48×		2−		7−	5−		13+	4
		3−			1−			
11+		3−		1−	4−	16×		
1−	7−	24×				15+		180×
			18×					
162×	2−	2÷	3−	5−	7−		14+	
					2÷	4−		14+
	9+	8−	63×			3		
5			4÷		3−		2÷	

SOLVING
TIME

13

+ − × ÷

1−		19+			15×	36×		
3÷	4−					3−		40×
	7+		18×		1−		18+	
12×	3−	3÷			14×	16×		
		22+	42×					4
			29+			1−		1−
19+	32×			15+			3÷	
			4−	3−	16+			11+
90×					2÷			

SOLVING TIME

+ − × ÷

40×			11+		13+		2−	3÷
3÷	2−		13+			6		
	144×		1−	14×		16+		20+
1−				5−				
	2÷		3−		7	48×		4−
5−	1−	8−		6+	2−		432×	
		3−	2−					
1−				9	4−	20×	11+	
5−		48×						4

SOLVING
TIME

+ − × ÷

8−	2÷		8+	14×			13+	
	1−	3−		5−		20+		
10×			48×		16+	10+		
	4−						7−	
1−		8+	192×		4−	25+		2−
3−	1							
	2−	6−		432×	45×		17+	
4−		8+	2−			4−	3−	
			21×					

SOLVING
TIME

+ − × ÷

7−	1−		189×		9+		11+	
	6−			3÷		16+	20+	
72×		3−	3−				2−	
10+			72×					
	5−		8+	567×		5	192×	
35×	8−	4÷				3÷		
			1−		1−		4−	
6×	2−	23+				7+		54×
		3÷			13+			

SOLVING TIME

+ − × ÷

2−		3−		11+	48×		2−	
10+		144×					3−	
70×			8+	28+				
	160×			3−	2÷		4−	
14+		8+			1−		12+	
		7	3÷	5−	7−	25+		
2	2−							
3÷		7−		20×	16+	35×		17+
2−		6−						

SOLVING TIME

+ − × ÷

5−	630×		36×	4−		6+	4−	
			2÷				96×	
3−		36×	36×		12+			2−
2÷				8−		168×		
20+		4−			54×		2÷	
11+		3÷	12+	40×			4÷	
	18+				72×		3−	
		4÷	2−			14+	6+	
1−			6−				7−	

SOLVING
TIME

+ − ✕ ÷

2÷	12✕			1−		2−		3÷
	19+			90✕	6−	11+	5−	
360✕								7−
2−		1−		72✕			640✕	
28✕		15+						1−
11+		5−		20+	14✕			
	11+	11+				11+		11+
			80✕		18+	126✕		
6	5−							

SOLVING
TIME

+ − × ÷

24×		15+			3−		2205×	
1	2−			3−		17+		
28×		7+		5−	2÷		14+	
2−	4−		6−					
	45×			18+			288×	
18+	108×	3÷			2÷			1
		3−		4	42×	18+	8+	
		35×						160×
48×			14+					

SOLVING
TIME

+ − × ÷

6×		800×		19+			4−	
2÷				1−		192×	3	17+
2÷	8−		21×	18×	3−			
	2÷					1−		
288×	252×			2−		14+	24×	
		1−	6	12+				
2−	1−		7+			2−	1−	
		31+	48×				4÷	
			2−			3	1−	

SOLVING
TIME

+ − × ÷

4−	8×		18+		11+		378×	1−
	54×		2−		19+			
1−	18+			14+			30×	
						7−	6+	
2−	2−		30×		15×		72×	
	4−	7+				2−	2−	
192×		13+	288×				1−	
	210×				2÷		5−	
			54×			20×		

SOLVING
TIME

+ − × ÷

108×			9+		504×	3÷		40×
2÷		63×		17+			17+	
3÷	3−	96×				5−		
			22+		15+			7−
2÷							4−	
2−		3−	2÷		4+			1−
1−			8+			3−	56×	
2÷	3×		60×	18×				42×
				216×				

SOLVING TIME

+ − × ÷

3÷	5−	270×		6	15+		4−	
				13+	2−		5−	3−
25+					18×			
3÷		24×			252×			
16+		16+			4−		5−	7+
	5−		16+		6×	2−		
1−	2−	2÷					22+	
		5−	17+			12+		
8+				13+		13+		

SOLVING
TIME

+ − × ÷

5−		480×		14+		6−		3−	
5−	2−			5−	144×		54×		
		4−			3				
20+	1−		1−		4÷	4−		6+	
		5−	42×			2÷			
5−				42×		10×		2÷	
40×			1−			144×			
84×				9	4−			4÷	
5	7−		1−			1−			

SOLVING
TIME

+ − × ÷

16+	1−		24×			2÷	11+	17+
	147×			17+				
	768×				21+			
		3−			378×		2−	
		192×				210×		10+
3−		5−		16+				
22+	16+				2÷	16+		
			5					15+
5−		6−			12+			

SOLVING TIME

+ − × ÷

12×			6+	20+		3−	3÷	4−
1920×				19+				
	40×					9×	2÷	18×
14+		54×		1−				
		2016×			13+		12+	
				10+			11+	
16+			15+	15+		13+	5−	
54×	105×						4÷	
			4÷		1−		4÷	

SOLVING
TIME

+ − × ÷

11+	3−		360×		5−	3÷	14+	
		63×	4÷					32×
15+	7			10+				
	4÷		17+		112×	9	2−	
	7+						14+	
	16+			20×		1−	20+	
1−	7−	3−	504×		15×			2
						2−		3÷
9+		9+		13+		30×		

SOLVING
TIME

+ − × ÷

27×			360×		2÷	2÷	13+	4−
2−		7+						
9+	5−		36×		2−		10+	
	15+		5−	2−	6	4−		
	21+				24×		7−	
		2−				7−		4−
24+	2−		10+		4÷		4−	
		5−		8−		13+		3÷
9		7−		21×				

SOLVING
TIME

+ − × ÷

135×			17+		4÷		22+	
10+		11+	14+		7−		60×	
2−				6+				
6+		17+		7+		16×		18+
10+			48×		17+	11+		
	2÷	11+	4−				16+	
				2÷	42×			2−
4÷		8+	162×			28×		
7+					2−		2÷	

SOLVING
TIME

+ − × ÷

28×	2−		270×	6−		11+		
	2÷			1−		45×	1−	
		15+		5−			5−	
40×			3÷	6×	1−		13+	2−
48×	1−				168×			
	13+		3−		8+		3÷	
	22+		1−				35×	2÷
3−		3−		126×	60×			
	5−					2	4−	

SOLVING
TIME

+ − × ÷

40×	31+				36×		3−	
		17+			60×		21×	
2−	2−							
	225×		2−	24×	8−	21×	2	10+
6−								
5−	48×	12+			2−	24×	126×	
			3−	105×				432×
11+		2−				1−		
			4÷		80×			

SOLVING
TIME

33

+ − × ÷

4÷		25+	96×	3÷		11+		210×
7+					6+			
16+	2÷		13+	8−	5−	7+	30×	
							144×	
	2−		5−	13+		288×		
	20+							3−
9+		3−		14+	7+		21×	
	378×					5−		2÷
8−		240×						

SOLVING
TIME

34

$+ - \times \div$

19+	27×		5−		12+	17+		
			1−				54×	
	19+		2−		5−			5−
		1−	3÷		27×	7−		
1−			3÷	40×		2−	18×	
17+	8+						3−	1
		12+			14+			20+
18×		1−		4	21+	12+		
8×			6					

SOLVING
TIME

+ − × ÷

4−		21+			6−		30×	
1−			135×	8×	54×		15+	1−
64×						15+		
	10+			180×				7−
20+		6−	10+			4÷		
17+			17+	5	16+	1−		56×
	3−					9+		
13+			864×	5−		15+	81×	

SOLVING
TIME

+ − × ÷

72×		14+		19+			16+	5
	11+		16+		12×	27+		
		7+					9	6×
27×	3−		2−				400×	
		96×		13+				
	6×				3−	31+		
1200×		3−						864×
21+			2÷		3÷			
		16+		4−		5+		

SOLVING TIME

+ − × ÷

45×		3	20+		13+	56×		
	9+					17280×		
3÷		11+		17+			20×	
	13+				5−			22+
3÷		11+		5−				
	22+	4−		19+	3−		4÷	
2÷					12+			6+
	16+	8−		42×		9	3−	
					40×			

SOLVING
TIME

+ − × ÷

23+		56×			2÷		15+	16+
1−		70×			18×			
	2÷		120×	864×				
22+	2−					20+		2÷
			4−			1120×		
112×		3÷	144×				432×	
	5		216×		22+			
8−				48×			360×	8
14×								

SOLVING TIME

+ − × ÷

3−		1−		48×	360×	8−	19+	
10+		16+						6−
5−			1−			2÷		
13+	8	24×		14×			20+	
			1−		7+		320×	
14+	5−		3÷	17+				
	13+	8−		15+				4÷
			9+		15×			
2	13+		10+		4−		2−	

SOLVING
TIME

+ − × ÷

120×	8×	2−		162×		7+		2−
		2−			112×		180×	
	8−		20+					2÷
14×		5−		315×	1−			
54×					12×		1−	
		210×		4÷			17+	
2−		9		16+				13+
18+	4	24×		13+	189×	13+		

SOLVING
TIME

+ − × ÷

2÷		240×	84×		13+	3−		21+
800×								
		24×		19+	19+		1−	1−
	15+	3÷				1−		
		2÷	3−				42×	
504×				210×		60×		
		23+	1−				24×	
18×			16+		8×	8	17+	
3−							8	

SOLVING
TIME

+ − × ÷

36×		7+	21+			90×		
1−			84×		11+		12×	
	3	2−			4−			270×
4÷		1−		15+		81×	28×	
16+		6+						
	270×	18×	15×		10+		24×	4−
						2−		
1−		48×	15+		90×		11+	
7−						2−		

SOLVING TIME

+ − × ÷

56×	20×			20+		13+		
	2÷		2−			108×		2−
		21×	7	12+	1−	2−		
1−							7−	
168×			11+		4÷	360×		15×
1−		360×	15+					
1−				3÷		5−	5−	10+
	216×	22+	84×					
						1	1−	

SOLVING
TIME

+ − × ÷

8×	405×		2÷		24+		13+
			3−		3×		
9+		336×	5−			144×	11+
3−				4−			
	4−		18+		8×		5−
7−		2÷		1344×		315×	
1−					1−	7+	162×
12+		2÷	5−	18×			
2÷					1−		4−

SOLVING
TIME

+ − × ÷

16+	81×		7+		1−		1−	
	2		6−		15+		2−	
	1512×		1−		17+		14×	4
		2÷	3÷					
240×			13+	2−		1−	4−	3÷
				8	6−			
15×			3÷			2÷		22+
42×	20+			9+	2÷		1−	
		48×			3÷			

SOLVING
TIME

+ − × ÷

1−		10×	6−		120×	12+		
1−			1−			2−	3−	13+
12+			18×	14+				
	7+					13+	7−	
12+		1−	17+		11+		5−	
5−							140×	8−
4÷	17+	2÷		2−	9+			
		3÷			4−			3−
5		24×		216×				

SOLVING TIME

+ − × ÷

6×		2−		4−		21+	16+	
4−		288×		3÷				10×
	2−		2−		19+	5−		
14+		6+				8−		15+
			189×		5	1−	4÷	
18+		56×		60×	3−			3÷
	16+					1−		
63×		90×	112×		1−		16+	
						4−		

SOLVING TIME

+ − × ÷

7	22+			42×		17+		
24×	1−	75×			756×	7−		7−
						13+		
5−		2−		3÷		16+		11+
270×		3−			4÷			
14+		56×	1−				12+	
				5−		3−		
8−		3÷	9+	20×		168×		
7−				5	1−		54×	

SOLVING
TIME

+ − × ÷

80×			12+		17+			432×
8+		10+			90×			
	3÷	4÷		9+	3024×			
19+		18×			3−	189×		
		1−	5+					
13+			315×		3	15+	10+	
56×	16+		16+					
		441×		5−	120×		24×	9+

SOLVING
TIME

+ − × ÷

42×		3−	6+		3−		56×	17+
15+			2−		20×			
		3−		23+		4−	2÷	8+
8+			2−					
	4÷		5−		189×	20+	2−	
45×	4÷		56×	3−				4−
	20+					4−	8+	
		63×	4÷	36×				
2−					2÷		45×	

SOLVING
TIME

+ −

9+			19+	14+			23+	
2−	22+			5−		5−		
				9+	22+		13+	5−
12+	6+	4+	6			12+		
				5−			11+	
	12+	2−	14+	23+				3
3−				14+	10+	2−	3−	
	20+		1−				3	8−
2−						5+		

SOLVING
TIME

+ −

2−	1−	19+			3−		7+	
		16+			19+		2−	
16+	13+			15+		20+		1
	9+					1−		6+
13+		7+	16+	16+			6−	
12+	7				22+	3+		11+
	21+	15+					11+	
				5−		3		16+
1		5−			13+			

SOLVING TIME [ ]

+ −

13+	9+	4−	9	20+	5−		17+	6+
					8+			
		5+	9+		13+		15+	15+
17+			2−		6+	17+		
	15+	13+	5−					14+
16+			10+			1−		
		3−		18+	12+	7+	2−	
4−	12+							10+
		10+		6−		5+		

SOLVING
TIME

+ −

6	2−		17+	18+	6+		11+	
2−	18+					4−		
		8+		7	19+			3−
7+		8−		6−		10+		
	7+		15+	2−		4	19+	
2−				6+			9	
12+	5−	9+		14+		2−		8
		4−	13+	12+		4−		7−
17+						8+		

SOLVING
TIME

+ −

3−	13+			7−		20+		
	7−	16+		6−		5−		3−
8+		14+		13+	2−	4+		
	1−		1−			22+	9+	
14+		23+		3−			8	11+
	21+			5−			4−	
		6+	4+	14+	5+	1		
14+						23+		
	15+				18+			

SOLVING TIME

+ −

12+		2−	11+	7	9+	5+		8+
14+	7+					4−		
			2−		22+		13+	
11+	10+		17+			3+		9+
	4−			13+	4+	13+		
2−		7−	9				15+	
5−			17+		1−	11+		6+
11+	11+	7	3−			18+		
				5−			1−	

SOLVING TIME

+ −

6−	25+	13+	5+	5−	16+	10+		3−
						3−	9+	
		2−	1−					15+
3	2−		9+	4−	4−			
4−	4−	8	8−				19+	
	2−		6	15+				13+
16+		20+	1−		9+			
1−	3−			7+	6−	7+		
	16+	7−					2	

SOLVING
TIME

+ −

5+		2−	12+	19+		22+		
15+						14+	5−	
	6	4−	3−	12+	2−		4−	2−
11+								
12+	16+	11+	7	1−	2−	5+		5−
						3−	8+	
	12+		10+		5			8+
6−	5−		1−		16+	6		
	3−		1−				8−	

SOLVING TIME

+ −

11+		15+		5+	16+			17+
6	17+		1−			9+		
17+				5	15+		3−	
		1−		13+		9+	15+	
17+		9+	4	5−			5−	
1−	11+		22+		7−	16+		1−
		16+		5−		9+		
10+	1				16+		7+	1−
	17+			12+				

SOLVING TIME

+ −

24+			7+			6−	16+	
23+	4−	6−	8+	4−			8+	
				10+	5−			
	4−	19+			12+			3+
16+			30+				9+	
			10+	5	5−	7−		13+
2−	14+						3−	
	5+		13+		16+	11+		23+
4−		6+						

SOLVING TIME _____

+ −

16+			6−		10+	11+		16+
8	8+	14+				14+	15+	
8+		2−	7−					
			8+			4−		14+
20+			3−		15+		3+	
10+		14+		17+		9+		
8−	5−		3−					14+
	11+	13+			17+	16+		
		6	14+				4+	

SOLVING
TIME

+ −

6	4−		5+		16+	4−		6+
10+	13+		6−			9+		
	9+	9+		5	5−		8+	
1−			6	11+	8−			1−
	8−	9+			2−	18+	2	
12+			18+					5−
	22+				5+	8+		
	3−		8−			4−	2−	3−
5−		17+			4			

SOLVING TIME

+ −

3−	2−	5−		1−	4+	5−	7+	4
		11+						
14+	15+	5+		14+	2−		12+	
		4−			17+	13+		
			11+			7	6−	
8+		22+		1−		22+		
16+		11+			20+	9+	4	
			12+				18+	
8+			4−			2−		

SOLVING TIME

+ −

14+		8+	6−	1−	1−		11+	
6−	15+				12+			
		5	1−		23+		6−	9+
13+		14+		20+				
6			2−			18+		
15+	3	9+		17+	4+		2−	
	14+	22+					9+	1−
				7−	4−			
8+		2−			14+		1−	

SOLVING TIME

65   **+ −**

1−		12+	23+	12+	6+		6−	
11+					18+	3−		1
9+				17+			10+	
12+		7−				3−	2−	4−
9+		2−		9+				
10+		15+			5−		18+	
3−		4+	4	3−		1−		
14+			12+	4	7−	3−	12+	
8	7+						14+	

SOLVING
TIME

+ −

8+		2−	15+		15+		11+	14+
15+	10+		1−		11+			
		2−	1−	7	7−			
6−	14+			11+			1−	
		2−			7+	1−		8−
4−	3−	14+		24+		16+		
		6−	8+				5−	
6+					7	14+	14+	
4−		1−		4−				

SOLVING
TIME

+ −

23+		1−		12+	9	15+		5−
	15+							
13+	4+	1−		5−	13+	9+	17+	
		6+	2−					12+
8+				13+		23+		
	24+		12+				7+	
2−			3	12+		12+		
9+		18+		11+	4−		6−	
	14+					9	6−	

SOLVING
TIME

+ −

23+		16+		6−		5−	13+	
			9	8+			16+	
5−	1−		10+					18+
	12+		12+		2	17+		
22+				1	7+	7−		
4−		5+	3−	15+		6+		4+
	7−				23+	12+		
2−		16+		3−		16+		
	12+						5−	

SOLVING
TIME

+ −

1−		17+	17+		3−	7−	16+	
16+							2−	
14+	8−	2−	7−		15+	2		
			11+			12+	2−	
	10+		1−	2−	3+		3−	
21+	3−						10+	
	10+		3−	7−		29+	10+	
	3−			3				
	4−		13+		4−		6−	

SOLVING
TIME

+ −

14+		2−	19+	20+			6+	
				15+		5+		2−
11+	5−			5+	12+		16+	
		7+			11+	4−		9+
18+	4−	3−						
			18+		9+		8−	7−
	1−	10+	17+		7			
11+					3+		3	17+
	3+		4−		15+			

SOLVING
TIME

+ −

7−	3−	3+	18+				18+	
			5−	5	26+			2−
9+	3−	4−		1−	2−		11+	
			6+			6		
17+				1−		6+		
16+			9	9+	11+		2−	
2−		16+					14+	2
3−			23+	5−		5+		17+
3−				5−				

SOLVING TIME

+ −

4−	1−	3−	17+		5−		5−	
			10+	5−	1−		4−	
17+	7−				4−		7+	5−
		8+		4+	3−	11+		
2−	13+						3−	
	20+			9+			8−	
16+	2−	12+	20+		11+	3−		2
			2−				4−	2−
1−		9		19+				

SOLVING TIME [   ]

+ −

6+		13+		20+		3−	3−	16+
8−	1−		7−		4−			
	8+					21+	6	
2−		2−		4−			1−	3+
2−		14+	13+			8+		
7−					3		2−	
5−	5+		14+		19+		12+	
	15+	6−				3	15+	
3		17+				10+		

SOLVING
TIME

+ −

4−		5−		3−	22+		3−	5+
3−	8+	30+			13+			
				3−		4+	2−	11+
15+		21+	9+		6+			
4	8−			5−			5−	
2−					12+		8+	17+
	6+			12+	13+	8		
1−		8+				12+	9+	
2−			4−				14+	

SOLVING
TIME

+ −

6+		14+		5−	2−	3−		12+
7−		2−				10+		
1−	16+		14+				19+	13+
			14+	3+				
6−		7+		13+		13+		17+
26+				1−				
	10+		21+		10+		3−	6−
11+	20+			7−		6		
						2−		3

SOLVING
TIME

+

20+			9+	8+		16+	16+	
	13+			6+	9+		18+	
15+	7+		8					
		18+			8+		11+	
13+	4+	6	14+	8+		11+		11+
		9+		12+			7	
10+	20+		7+		13+		7+	
			9+	9+		12+		9+
	9+			11+		12+		

SOLVING TIME

+

13+			10+	4+	18+			17+
20+		2			5+	11+	14+	
	18+		9+					
16+		4+		19+	15+		5+	15+
	10+	20+			24+			
							10+	
14+	5+	16+	8+	11+				16+
				15+	5+			
	17+					19+		

SOLVING
TIME

+

13+	1	9+	7+	12+	9+	18+	10+	
							9+	16+
18+	14+		6	6+	11+			
	17+		14+				14+	
		17+		5+		7+		
4+			13+		23+			5+
13+	15+	5+		13+		21+		
		2	17+					9+
	9+			14+			9	

SOLVING
TIME

+

27+	9+		7+		25+			7+
		10+		14+	17+		4+	
	10+		13+			7+		
7+	8+			13+			12+	16+
	5+	14+				12+		
20+		4+	17+	12+			9+	6
					21+			8+
11+		5+				10+	24+	
8+		8+		5+				

SOLVING
TIME

+

9+	9+		14+	13+	5+		12+	
	12+				11+		7+	
9+		18+	7+	12+			16+	
7+	11+			16+			5+	
			10+		24+		7+	
15+		6+				9	6+	
19+	9+	13+		3+	4+		6	12+
		8+	20+		23+			
					7		11+	

SOLVING
TIME

+

7+	19+			7	6+		19+	
	16+	13+		11+		13+		
6+		18+			6+		8+	
		1			14+		16+	
19+			18+			8+		
9+		11+		19+	13+	4	19+	7+
9+		19+						
11+	11+				11+			8+
		8+			21+			

SOLVING TIME

+

13+		7+		9+		22+		
8+		13+	9+	3+	18+		17+	
7+	17+							15+
		12+	10+	23+		9+	9+	
11+								
	6+		9+		14+		12+	11+
		17+		7+	16+	8+		
16+			17+				7+	
13+				14+		6+		

SOLVING
TIME

+

14+			20+	18+			1	9+
10+	11+	14+		20+		7+		
							13+	16+
	11+	4+	17+		15+			
30+				3+			13+	
	9	18+		9+		5+	10+	
				13+			9+	
3+		11+	9+	8+		20+		
11+				16+			8+	

SOLVING
TIME

+

16+		11+	7+	17+	17+	11+		9+
9+						3+		
14+	7+		11+				6+	16+
	9+	9+		13+		13+		
		11+	21+				9+	12+
13+	5+		6+		9+			
		7+	15+	6+		12+		
13+				7+	8+	13+	11+	
	16+						13+	

SOLVING
TIME

85

+

9+		17+	16+		8+	10+	12+	
			7+				18+	
7+		25+		14+		12+		
17+				16+	10+		11+	7+
	6+	12+			10+			
		7+			24+		11+	
23+	11+		6+		10+			
		11+	8+			11+		23+
13+				3+				

SOLVING
TIME

+

6+	13+		12+	20+		16+	11+	9+
				13+				
8+	7+		17+		12+		7+	
	7+	16+				11+		13+
16+				7	13+		5+	
	11+		15+	4+				15+
7+		15+		5+		13+		
12+	15+		14+		10+	27+		3+

SOLVING TIME

+

12+	11+			12+		13+	24+	2
	6+	19+	10+					
				6+	10+	15+	13+	9+
17+	6+	9+	12+					
				3	5+	10+	7+	15+
7+	15+	14+	3+	16+				
					8+		7+	12+
14+		9+	23+					
7			13+		11+			

SOLVING TIME

+

6+	22+	10+	15+		9+	13+	11+	
			3				26+	
		17+			3+			
17+	9+	8+		26+	10+		9+	7+
					8+			
	21+		12+	10+		12+		
8		14+			14+	9+		
5+			10+			10+		8
	8+		11+		10+		14+	

SOLVING
TIME

+

15+		10+		10+		6+		13+
17+		12+	11+		15+			
7			15+		12+			8+
12+				19+	17+	8+		
	25+		11+			3		12+
14+					5+	16+		
	18+						21+	6
5+	11+		9+					
	5+		9+		15+		13+	

SOLVING TIME

+

12+	14+	20+			8+		5+	
		9+	10+	19+	16+	19+		
7+						18+	4+	
	7+		6+					15+
12+	13+			9+		5+		
	20+	10+	22+	3+		13+		18+
				2	5+	11+		
7+	4+			25+			11+	15+
	11+							

SOLVING
TIME

+

13+		18+		14+		18+	13+	
5+	12+			5				
		21+			18+	9+	9+	10+
9+		13+						
8+			6+	12+		12+		9+
17+	13+			4+		17+		
	9+	14+		8+			16+	
5+			13+		8+		5+	
	10+		21+				11+	

SOLVING
TIME

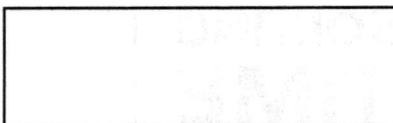

+

6+		15+		8+		17+		18+
13+		19+			12+			
	5+		16+			11+		5+
16+				23+	13+			
11+			11+			8+		14+
8+	5+	19+		7+			19+	
				13+	8+	13+		
13+	19+		7+				9+	
				13+		1	10+	

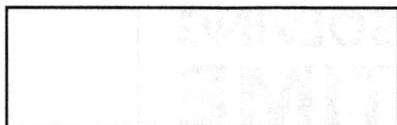

SOLVING TIME

+

18+			14+		8+	5+		2
10+		4+		18+		15+		22+
	9	14+					14+	
7+		22+	13+			11+		
12+			19+					
	10+		14+	8+		5	19+	
16+		15+			14+			13+
			9+		10+			
	9+			6+		20+		

SOLVING TIME

+

8+		13+		18+	4+		20+	
7+	9+	13+	8+		25+	6+		14+
							16+	
14+			29+					6+
3+	15+					5+		
	14+		19+			18+	10+	
12+		10+		8+	12+		3+	16+
	21+							
		9+			11+		9+	

SOLVING
TIME

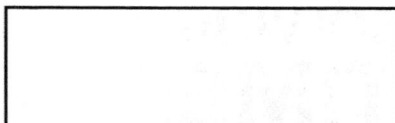

+

7+			8+		15+		16+	16+
15+		18+		12+				
	24+				11+		5+	
6+	12+			17+	8	16+		
		7+			6+		11+	8+
20+	6+		7		21+			
		16+		11+				18+
10+	19+		4+					
		15+		9+		11+		

SOLVING TIME

+

8+		11+		9+		8+		12+
20+	3+		7+	10+	7+	16+		
		13+				15+		20+
9+			3+	18+				
24+	8+			10+		9+		
		20+			3	7+		
10+	4+	19+		11+	8+		13+	3+
		17+			14+			
3			8+		25+			

SOLVING TIME

+

11+	14+		15+		6+	3+	15+	
		14+	16+	12+				11+
21+					9+	11+		
	12+	14+				7+		2
		13+		17+			7+	8+
3+	11+		10+	20+		13+		
				16+			15+	28+
12+	9+	8+				9		
		4		9+				

SOLVING
TIME

+

9+	7+	6+	10+		15+		5+	
			8+		17+	13+		18+
9	15+			18+		7+	17+	
8+	23+	7+	14+					
					7+	13+		6+
11+		17+	17+					
			5+		20+			
	15+			11+		7+	16+	11+
12+		8+		3+				

SOLVING TIME

+

8+	18+		7+	10+		22+		6+
	6+			15+				
6+		14+	4+		16+	8+	18+	8
			10+	14+				
1	7+				9+	11+	16+	
21+	7+	9+	14+				8	
			12+	16+			11+	
	26+	12+		14+	3+		4+	8+
						6		

SOLVING TIME

+

11+		20+		9+	9+	9+	4+	
10+		10+					10+	
14+	15+		13+		14+	18+		5+
		10+	6+					
14+			18+			10+	5	21+
	11+		15+		22+			
		1	9+				16+	
11+		11+		11+			7+	
7		8+			8+		13+	

SOLVING
TIME

Bonus No-Op Puzzle! This puzzle does not include operations after the target numbers...
you must deduce them! Note that the operation for cages with 3 or more squares must be
addition or multiplication.

101

+ − ✕ ÷

63		252	4		20			10
			18		128			
3		21				80		
4	5			13	72		5	72
	54	1				1		
		7		2			11	1
200		17	2	5	2			
					3	16	8	
40			72				2	

SOLVING
TIME

Bonus No-Op Puzzle! This puzzle does not include operations after the target numbers...
you must deduce them! Note that the operation for cages with 3 or more squares must be
addition or multiplication.

102

+ − × ÷

108	3		16			63		
		16	42		7			5
56			2	2	405			120
1					12			
	2		18				24	
11		3		7	63		72	
3	4	15			36			
		18		2	2		6	15
9	2				4			

SOLVING
TIME

Bonus No-Op Puzzle! This puzzle does not include operations after the target numbers... you must deduce them! Note that the operation for cages with 3 or more squares must be addition or multiplication.

+ − ✕ ÷

16		6		45	4	3		19
	2					3	11	
12		2		12		315		
		8	32					
280			24		8	2	2	180
	6	2		288				
16	20					11		
		105		13		7		2
10				13		144		

SOLVING
TIME

Bonus No-Op Puzzle! This puzzle does not include operations after the target numbers...
you must deduce them! Note that the operation for cages with 3 or more squares must be
addition or multiplication.

104

+ − × ÷

10		1		1	10	160		
	4		2			81	2	
32	3			5			21	5
		21			24	15		
12	13	18		7				24
		288		20	1			
2					6		10	
15	18			11	25			4
		2					3	

SOLVING
TIME

Bonus No-Op Puzzle! This puzzle does not include operations after the target numbers... you must deduce them! Note that the operation for cages with 3 or more squares must be addition or multiplication.

105

＋ － ✕ ÷

960		23		24	6			1
14						5		
	60		180			20		
	3		11		4		2	
2		42				23	105	2
	10		3					
13		8		3		30	40	
147			16				4	
				4		19		

SOLVING TIME

SOLUTIONS

1

8× 8	**11+** 2	3	**17+** 9	**24×** 6	**3−** 4	7	**4−** 5	1
1	6	**3÷** 2	5	4	**1−** 7	**243×** 3	9	**2÷** 8
2− 7	5	6	3	**18×** 2	8	9	1	4
15+ 4	**4−** 7	**7+** 5	2	9	**3÷** 6	**96×** 1	8	3
5	3	**8×** 1	8	**6−** 7	2	4	**108×** 6	9
6	**2÷** 4	**22+** 8	7	1	**135×** 9	5	3	2
18× 9	8	7	**12+** 6	**3** 3	**12+** 1	**11+** 2	4	5
2	**22+** 9	4	1	5	3	8	**42×** 7	6
3÷ 3	1	9	**2÷** 4	8	**1−** 5	6	2	7

2

3÷ 6	2	**1−** 8	**3−** 3	**3−** 5	**4−** 1	**196×** 7	4	**5−** 9
13+ 2	**15+** 3	9	6	8	5	**3−** 1	7	4
8	5	**4−** 6	2	**21×** 7	3	4	**405×** 9	**8×** 1
3	7	**40×** 1	**2÷** 4	2	6	**2−** 9	5	8
42× 7	1	5	8	**19+** 9	4	**5+** 3	2	**3÷** 6
18+ 4	6	**12+** 7	5	1	9	**21+** 8	**10+** 3	2
5	9	**12×** 4	**8+** 7	**48×** 6	8	2	1	**23+** 3
18+ 9	8	3	1	**1−** 4	**9+** 2	5	6	7
1	**2÷** 4	2	**9** 9	3	7	6	8	5

3

15× 3	5	**9+** 7	**168×** 4	6	**4÷** 2	8	**10+** 9	1
3÷ 1	3	2	**21+** 8	7	**120×** 5	4	6	**13+** 9
5 5	**168×** 7	6	9	4	**18+** 8	**84×** 2	1	3
4÷ 8	4	**4−** 3	7	9	1	6	**7+** 2	5
2	**7−** 9	**1−** 5	**8+** 1	3	4	7	**5−** 8	**2−** 6
16+ 7	2	4	**1−** 6	5	9	1	3	8
9	**16+** 6	1	**4÷** 2	8	**1−** 7	**15×** 3	**35×** 5	**2÷** 4
24× 4	8	9	3	**2÷** 1	6	5	7	2
6	1	8	5	2	**6−** 3	9	**3−** 4	7

4

2÷ 4	**18×** 9	2	**3÷** 3	1	**2−** 8	6	5	**13+** 7
2	**2−** 6	4	**245×** 7	**90×** 9	5	**144×** 3	1	**7−** 8
13+ 9	4	7	5	2	**3** 3	8	6	1
40× 8	**3÷** 1	**1−** 3	2	**5−** 4	9	**2−** 5	7	**20+** 6
5	3	**10+** 6	4	**7−** 8	1	**13+** 7	2	9
1	**42×** 7	**22+** 9	8	3	6	2	4	5
3	2	5	**30×** 1	6	**252×** 7	9	**17+** 8	**12+** 4
26+ 7	8	**48×** 1	6	5	**22+** 2	4	9	3
6	5	8	9	7	4	**1** 1	3	2

5

180× 4	5	9	14+ 7	4÷ 2	8	13+ 1	3	16+ 6
30× 2	84× 3	6	1	8	4	3− 5	7	36× 9
3	7	2÷ 1	13+ 6	5	2	8	1− 9	4
5	4	2	378× 9	7	5− 1	6	8	3÷ 3
16+ 7	2	5− 8	3	6	9	13+ 4	30× 5	1
1	216× 9	14× 7	2	20× 4	5	6× 3	6	14+ 8
6	8	3	160× 4	1134× 9	7	2	1	5
17+ 9	11+ 1	5	8	3	6	3− 7	4	5− 2
8	6	4	6+ 5	1	3	14+ 9	2	7

6

6+ 1	5	3÷ 2	1944× 8	3	9	168× 4	6	7
144× 3	8	6	9	4÷ 4	1	7− 2	2− 7	5
6	3÷ 1	3	14× 7	18+ 2	8	9	180× 5	4
11+ 4	7	20× 1	2	5	3	15+ 6	9	5− 8
4÷ 2	20+ 9	4	5	13+ 6	7 7	1	8	3
8	6	5	4÷ 1	7	1− 4	3	2÷ 2	7− 9
16+ 5	3	8	4	9 9	13+ 6	7	1	2
2− 7	7− 2	9	3÷ 3	1	5	13+ 8	1− 4	5− 6
9	3− 4	7	2− 6	8	2	7+ 5	3	1

7

16× 8	2	1	2÷ 6	3	19+ 9	140× 5	3− 4	7
108× 2	13+ 3	5	1− 8	5− 6	7	4	8− 1	9
6	5	4÷ 8	9	1	3	7	3÷ 2	24× 4
9	7 7	2	4− 1	5	24+ 8	4	6	3
30+ 1	8	84× 6	7	1− 4	5	3	2− 9	2
5	48× 4	3	2	15+ 8	5− 6	9	7	24× 1
7	9	4	5	2	1	21+ 6	3	8
3÷ 3	1	36× 9	4	189× 7	8	2	5	240× 6
17+ 4	6	7	3	9	2÷ 2	1	8	5

8

1− 6	7	10+ 9	1	5− 3	5	4− 8	64× 4	2
5− 1	6	28× 7	4	8	9	3÷ 3	10+ 2	5
21+ 4	9	8 8	3− 2	5	168× 7	1	3	1− 6
80× 2	8	20+ 5	9	6	3	4 4	6× 1	7
5	8× 4	1	4− 3	7	8	18+ 2	6	6− 9
8	17+ 5	2	48× 6	24× 4	1 1	9	2− 7	3
9	2	72× 3	8	1	6	7	5	2÷ 4
13+ 3	1	6	7	2− 2	4	4− 5	9	8
7	3	4	5	108× 9	2	6	7− 8	1

9

3 3	40× 8	5	10+ 7	1	17+ 6	2	9	140× 4
10+ 1	9	4– 4	2	4– 3	7	2– 8	6	5
4– 6	2	8	180× 1	5	9	4	7+ 3	7
56× 8	7	20+ 6	6– 9	2 2	2– 3	5	4	1
11+ 4	5	9	3	21+ 7	2÷ 8	10+ 1	4– 2	6
5	84× 3	7	6	8	4	9	12+ 1	2
2	4	2÷ 1	160× 8	19+ 6	10× 5	4– 3	7	9
2– 9	10+ 6	2	5	4	1	1– 7	8	24× 3
7	1	3	4	9	2	6	11+ 5	8

10

2÷ 4	30× 3	2	22+ 5	7	6	9 9	5– 1	7– 8
2	1– 8	7+ 4	3	9	2– 5	7	6	1
15+ 8	9	28× 7	4	1– 2	1– 1	60× 3	5	13+ 6
6	21+ 5	9	24× 1	3	2	4	8 8	7
1	7	3	8	17+ 6	15+ 4	4÷ 2	4– 9	5
3– 5	2	1	6	4	3	8	63× 7	9
13+ 9	4	180× 6	8+ 7	1	8	7+ 5	2	3 3
168× 3	6	7– 5	3– 9	8+ 8	7	2– 1	4	2
7	1	8	2	15+ 5	9	6	12× 3	4

11

7– 1	8	48× 2	3	5 5	24× 4	6	26+ 9	7
80× 2	5	8	3÷ 1	3	9	19+ 7	6	4
8	2– 9	7	35× 5	6– 2	2÷ 1	3	20× 4	3– 6
15× 3	1– 6	2– 4	7	8	2	5	1	9
5	7	6	22+ 9	12× 1	3	4	360× 8	2÷ 2
5– 7	2	3	4	6	1– 8	9	5	1
2– 4	15× 3	5	3– 6	9	7	1	48× 2	8
6	1	7– 9	2	20× 4	5	1– 8	7	3
5– 9	4	16+ 1	8	7	3÷ 6	2	2– 3	5

12

48× 1	6	2– 7	5	7– 2	5– 3	8	13+ 9	4 4
2	4	3– 5	8	9	1– 6	7	3	1
11+ 4	7	3– 6	3	1– 5	4– 9	16× 1	2	8
1– 8	7– 2	24× 3	1	4	5	15+ 6	7	180× 9
7	9	8	18× 6	3	1	2	4	5
162× 3	2– 5	2÷ 4	3– 7	5– 1	7– 2	9	14+ 8	6
9	3	2	4	6	2÷ 8	4– 5	1	14+ 7
6	9+ 8	8– 1	63× 9	7	4	3 3	5	2
5 5	1	9	4÷ 2	8	3– 7	4	2÷ 6	3

13

1-7	6	19+5	8	2	15×3	36×4	1	9
3÷2	3	4-7	4	5	1	3-6	9	40×8
6	7+4	3	18×2	1	8	1-9	18+7	5
12×4	3-8	3÷1	3	9	7	14×2	16×5	6
3	5	22+9	42×6	7	2	1	8	4-4
1	7	6	29+9	8	5	1-3	4	1-2
19+9	32×1	8	7	15+4	6	5	3÷2	3
8	2	4	4-5	3-3	16+9	7	6	11+1
90×5	9	2	1	6	2÷4	8	3	7

14

40×1	5	8	11+7	4	3	13+9	2-2	3÷6
3÷3	2-9	7	13+5	8	1	6-6	4	2
9	144×4	6	1-3	14×7	2	1	16+8	20+5
1-4	6	1	2	5-3	8	7	5	9
5	2÷8	4	3-9	6	7-7	48×2	1	4-3
5-2	1-3	8-9	1	6+5	2-4	8	432×6	7
7	2	3-5	2-4	1	6	3	9	8
1-8	7	2	9-6	9	4-5	20×4	11+3	1
5-6	1	48×3	8	2	9	5	7	4-4

15

8-9	2÷4	8	8+3	14×1	2	7	13+6	5
1	1-9	3-6	5	5-3	8	20+4	7	2
10×5	8	9	48×4	2	7	16+1	10+3	6
2	4-7	3	6	5	4	9	7-8	1
1-7	6	8+2	192×8	4	1	25+3	5	2-9
3-3	1-1	4	2	6	5	8	9	7
6	2-3	6-7	1	432×8	45×9	5	17+2	4
4-8	5	8+1	7	9	6	4-2	3-4	3
4	2	5	9	21×7	3	6	1	8

16

7-8	1-6	5	189×7	3	9+1	4	11+9	2
1	6-8	2	9	3÷6	4	16+3	20+7	5
72×9	4	3-3	3-8	2	6	7	2-5	1
10+4	2	6	5	72×1	8	9	3	7
6	5-3	8	8+1	567×9	7	5-5	192×2	4
35×7	8-1	4÷4	2	5	9	3÷6	8	3
5	9	1	1-6	7	3	1-2	4-4	8
6×3	2-5	23+7	4	8	2	7+1	6	54×9
2	7	3÷9	3	4	13+5	8	1	6

17

2−9	7	3−8	5	11+1	48×2	3	2−6	4
10+1	9	144×4	6	3	7	8	3−5	2
70×5	2	6	8+4	28+8	9	1	3	7
7	160×8	3	1	3−6	2÷4	2	4−9	5
14+8	4	8+1	7	9	1−5	6	12+2	3
6	5	7·7	3÷3	5−2	7−8	25+9	4	1
2·2	2−3	5	9	7	1	4	8	6
3÷3	1	7−9	2	20×4	16+6	35×5	7	17+8
2−4	6	6−2	8	5	3	7	1	9

18

5−6	630×2	5	36×9	4−4	8	6+1	4−7	3
1	9	7	4	2÷3	6	5	96×8	2
3−5	8	36×9	36×3	2	12+1	4	6	2−7
2÷2	1	4	6	8−9	7	168×8	3	5
20+9	5	4−6	2	1	54×3	7	2÷4	8
11+8	6	3÷3	12+7	40×5	2	9	4÷1	4
3	18+7	1	5	8	72×4	2	3−9	6
7	4	4÷2	8	6	9	14+3	6+5	1
1−4	3	8	6−1	7	5	6	7−2	9

19

2÷4	12×1	2	6	1−7	8	2−5	3	3÷9
2	19+8	4	7	90×5	6−1	11+6	5−9	3
360×8	9	5	3	6	7	2	4	7−1
2−9	7	1−1	2	72×4	6	3	640×5	8
28×7	4	6	9	1	3	8	2	1−5
11+1	5	5−9	20+4	3	14×2	7	8	6
5	11+2	11+3	1	8	9	11+4	6	11+7
3	6	7	80×8	2	18+5	126×9	1	4
6·6	5−3	8	5	9	4	1	7	2

20

24×8	3	15+2	4	1	3−9	6	2205×7	5
1·1	2−4	6	8	3−2	5	17+9	3	7
28×4	7	7+1	6	5−8	2÷3	5	14+2	9
2−5	4−8	4	6−7	3	6	1	9	2
3	45×5	9	1	18+7	4	2	288×8	6
18+7	108×2	3÷3	9	5	2÷8	4	6	1·1
9	6	3−5	2	4·4	42×7	18+8	8+1	3
2	9	35×7	5	6	1	3	4	160×8
48×6	1	8	14+3	9	2	7	5	4

6× 6	1	800× 5	4	19+ 8	9	2	4− 7	3
2÷ 1	2	8	5	1− 7	6	192× 4	3 3	17+ 9
2÷ 4	8− 9	1	21× 3	18× 2	5	3− 6	8	7
2	2÷ 3	6	7	9	8	1− 5	4	1
288× 8	252× 7	4	9	2− 5	3	14+ 1	24× 2	6
9	4	1− 3	6 6	12+ 1	7	8	5	2
2− 3	1− 5	2	7+ 1	6	4	2− 7	1− 9	8
5	6	31+ 7	48× 8	3	2	9	4÷ 1	4
7	8	9	2− 2	4	1	3 3	1− 6	5

4− 5	8× 8	1	18+ 4	6	11+ 9	2	378× 7	1− 3
1	54× 2	3	2− 5	8	19+ 7	6	9	4
1− 2	18+ 1	9	7	14+ 3	8	4	30× 5	6
3	9	6	2	7	4	7− 8	6+ 1	5
2− 7	2− 4	2	30× 6	5	3	15× 1	72× 8	9
9	4− 3	7+ 4	1	2	5	2− 7	2− 6	8
192× 6	7	13+ 5	288× 8	4	1	9	1− 3	2
4	210× 5	8	9	1	6	2÷ 3	5− 2	7
8	6	7	54× 3	9	2	20× 5	4	1

108× 6	9	2	9+ 5	4	504× 7	3÷ 3	1	40× 8
2÷ 2	4	63× 7	9	17+ 1	3	6	17+ 8	5
3÷ 3	3− 5	96× 8	7	9	4	5− 2	6	1
9	8	4	22+ 1	6	15+ 5	7	3	7− 2
2÷ 1	2	3	8	7	6	4	4− 5	9
2− 5	7	3− 6	2÷ 4	8	4+ 2	1	9	1− 3
1− 7	6	9	8+ 3	5	1	3− 8	56× 2	4
2÷ 8	3× 3	1	60× 6	18× 2	9	5	4	42× 7
4	1	5	2	216× 3	8	9	7	6

3÷ 3	5− 4	270× 9	2	6 6	15+ 8	7	5	1
1	9	5	3	13+ 7	2− 6	4	5− 8	3− 2
25+ 6	8	4	7	1	18× 9	2	3	5
3÷ 2	6	24× 3	8	5	252× 7	1	4	9
16+ 4	7	16+ 8	6	2	4− 5	9	5− 1	7+ 3
5	5− 2	7	9	16+ 3	1	6× 8	2− 6	4
1− 8	2− 5	2÷ 2	1	4	3	6	22+ 9	7
9	3	5− 1	4	17+ 8	2	12+ 5	7	6
8+ 7	1	6	5	13+ 9	4	13+ 3	2	8

25

8	3	2	6	5	9	1	7	4
1	6	5	8	2	4	9	3	7
6	8	1	5	7	3	4	2	9
4	7	6	2	3	8	5	9	1
7	9	3	1	6	2	8	4	5
9	4	8	7	1	6	2	5	3
2	5	4	9	8	7	3	1	6
3	1	7	4	9	5	6	8	2
5	2	9	3	4	1	7	6	8

26

2	7	6	1	8	3	4	5	9
4	1	7	3	9	2	8	6	5
6	4	8	7	5	1	2	9	3
3	8	5	2	1	7	9	4	6
1	3	4	8	6	9	5	7	2
8	5	9	4	7	6	3	2	1
5	9	1	6	2	8	7	3	4
9	6	2	5	3	4	1	8	7
7	2	3	9	4	5	6	1	8

27

1	2	6	5	8	4	7	9	3
5	6	8	1	9	2	4	3	7
8	5	2	7	3	6	1	4	9
7	4	3	6	5	1	9	8	2
4	1	9	3	6	8	2	7	5
2	8	7	4	1	9	3	5	6
3	9	4	2	7	5	8	6	1
6	7	1	9	4	3	5	2	8
9	3	5	8	2	7	6	1	4

28

2	9	6	5	8	1	3	4	7
4	5	7	2	9	6	1	3	8
6	7	9	8	3	2	5	1	4
3	4	1	6	2	8	9	7	5
5	3	4	9	1	7	2	8	6
1	6	3	7	5	4	8	2	9
8	1	5	4	6	3	7	9	2
9	8	2	3	7	5	4	6	1
7	2	8	1	4	9	6	5	3

29

27× 3	1	9	360× 5	8	2÷ 2	2÷ 4	13+ 6	4− 7
2− 4	6	7+ 5	2	9	1	8	7	3
9+ 6	5− 8	3	36× 9	4	2− 5	7	10+ 2	1
1	15+ 2	7	5− 8	2− 5	6 6	4− 9	3	4
2	21+ 9	6	3	7	24× 4	5	7− 1	8
5	7	2− 4	6	2	3	7− 1	8	4− 9
24+ 7	2− 3	1	10+ 4	6	4÷ 8	2	4− 9	5
8	4	5− 2	7	8− 1	9	13+ 3	5	3÷ 6
9 9	5	7− 8	1	21× 3	7	6	4	2

30

135× 9	3	5	17+ 8	2	4÷ 4	1	22+ 7	6
10+ 4	6	11+ 8	14+ 3	7	7− 2	9	60× 5	1
2− 7	9	3	6+ 4	5	1	2	6	8
6+ 1	5	17+ 6	7	7+ 4	3	16× 8	2	18+ 9
10+ 3	7	4	48× 6	8	9	17+ 5	11+ 1	2
2	2÷ 4	11+ 9	4− 5	1	8	6	16+ 3	7
5	8	2	1	2÷ 6	42× 7	4	9	2− 3
4÷ 8	2	8+ 1	162× 9	3	6	28× 7	4	5
7+ 6	1	7	2	9	2− 5	3	2÷ 8	4

31

28× 7	2− 5	3	270× 9	6− 8	2	11+ 4	6	1
4	2÷ 1	6	5	1− 7	8	45× 9	1− 3	2
1	2	15+ 7	6	5− 4	9	5	5− 8	3
40× 5	8	2	3÷ 3	6× 1	1− 7	6	13+ 4	2− 9
48× 2	4	5	1	6	168× 3	8	9	7
3	13+ 9	4	3− 8	5	8+ 1	7	3÷ 2	6
8	22+ 7	9	1− 2	3	6	1	35× 5	2÷ 4
3− 9	6	3− 1	4	126× 2	60× 5	3	7	8
6	5− 3	8	7	9	4	2 2	1	4− 5

32

40× 2	31+ 6	3	7	9	36× 4	1	3− 8	5
5	4	17+ 8	3	6	60× 2	9	21× 7	1
2− 8	2− 9	7	4	2	6	5	1	3
6	225× 5	9	2− 8	24× 3	8− 1	21× 7	2 2	10+ 4
6− 7	1	5	6	8	9	3	4	2
5− 4	48× 8	12+ 2	9	1	2− 5	24× 6	126× 3	7
9	3	1	3− 2	105× 5	7	4	6	432× 8
11+ 1	2	2− 4	5	7	1− 3	8	9	6
3	7	6	4÷ 1	4	80× 8	2	5	9

33

4÷ 8	2	25+ 1	96× 6	3÷ 3	9	11+ 7	4	210× 5
7+ 3	4	9	2	8	6+ 1	5	6	7
16+ 4	2÷ 3	8	13+ 9	8− 1	5− 7	7+ 6	30× 5	2
5	6	7	4	9	2	1	144× 8	3
6	2− 5	3	5− 8	13+ 7	4	288× 9	2	1
1	20+ 7	5	3	2	8	4	9	3− 6
9+ 7	8	3− 4	1	14+ 6	7+ 5	2	21× 3	9
2	378× 9	6	7	5	3	5− 8	1	2÷ 4
8− 9	1	240× 2	5	4	6	3	7	8

34

19+ 3	27× 1	9	5− 7	2	6	12+ 8	17+ 5	4
5	3	1	1− 8	7	2	4	54× 9	6
7	19+ 6	8	5	2− 3	4	5− 9	1	5− 2
4	5	1− 3	3÷ 2	6	27× 9	7− 1	8	7
1− 6	7	4	3÷ 3	40× 8	1	5	18× 2	9
17+ 8	8+ 4	2	9	5	3	7	3− 6	1 1
9	2	12+ 7	4	1	14+ 8	6	3	20+ 5
18× 2	9	1− 6	1	4 4	21+ 5	12+ 3	7	8
8× 1	8	5	6 6	9	7	2	4	3

35

4− 7	3	21+ 1	4	9	6− 8	2	30× 5	6
1− 3	2	7	135× 5	8× 8	54× 1	9	15+ 6	1− 4
64× 4	8	9	3	1	6	15+ 7	2	5
2	10+ 6	3	1	180× 4	5	8	7	7− 9
20+ 6	5	6− 8	10+ 7	3	9	4÷ 4	1	2
8	9	2	17+ 6	5 5	16+ 7	1− 3	4	56× 1
9	3− 4	5	2	6	3	9+ 1	8	7
13+ 5	1	4	864× 9	5− 7	2	15+ 6	81× 3	8
1	7	6	8	2	4	5	9	3

36

72× 2	9	14+ 3	6	19+ 8	7	4	16+ 1	5 5
4	11+ 1	5	16+ 2	9	12× 3	27+ 6	7	8
8	2	7+ 6	5	1	4	7	9 9	6× 3
27× 3	3− 4	1	2− 9	7	6	8	400× 5	2
9	7	96× 4	3	13+ 6	8	5	2	1
1	6× 3	2	8	4	3− 5	31+ 9	6	7
1200× 5	6	3− 7	4	3	2	1	8	864× 9
21+ 7	5	8	2÷ 1	2	3÷ 9	3	4	6
6	8	16+ 9	7	4− 5	1	5+ 2	3	4

37

45×		3	20+			13+		56×
1	**9**	**3**	**8**	**5**	**6**	**2**	**7**	**4**
5	**1**⁹⁺	**2**	**3**	**4**	**7**	**6**¹⁷²⁸⁰ˣ	**9**	**8**
2³÷	**6**	**4**¹¹⁺	**7**	**3**¹⁷⁺	**9**	**8**	**1**²⁰ˣ	**5**
6	**3**¹³⁺	**8**	**4**	**1**	**2**⁵⁻	**7**	**5**	**9**²²⁺
9³÷	**2**	**6**¹¹⁺	**5**	**8**⁵⁻	**3**	**1**	**4**	**7**
3	**7**²²⁺	**5**⁴⁻	**9**	**2**¹⁹⁺	**1**	**4**³⁻	**8**	**6**⁴÷
4²÷	**8**	**7**	**6**	**9**	**5**¹²⁺	**3**	**2**	**1**⁶⁺
8	**5**¹⁶⁺	**1**⁸⁻	**2**	**7**⁴²ˣ	**4**	**9**⁹	**6**	**3**³⁻
7	**4**	**9**	**1**	**6**	**8**⁴⁰ˣ	**5**	**3**	**2**

38

23+		56×			2÷		15+	16+
6	**9**	**8**	**7**	**1**	**4**	**2**	**5**	**3**
4¹⁻	**8**	**5**⁷⁰ˣ	**2**	**7**	**3**¹⁸ˣ	**1**	**9**	**6**
5	**4**²÷	**2**	**3**¹²⁰ˣ	**9**⁸⁶⁴ˣ	**8**	**6**	**1**	**7**
3²²⁺	**6**²⁻	**4**	**5**	**8**	**2**	**9**²⁰⁺	**7**	**1**²÷
7	**3**	**9**	**1**⁴⁻	**5**	**6**	**8**¹¹²⁰ˣ	**4**	**2**
2¹¹²ˣ	**7**	**1**³÷	**8**	**3**¹⁴⁴ˣ	**5**	**4**	**6**⁴³²ˣ	**9**
8	**5**⁵	**3**	**9**²¹⁶ˣ	**6**	**1**	**7**²²⁺	**2**	**4**
9⁸⁻	**1**	**6**	**4**⁴⁸ˣ	**2**	**7**	**5**	**3**³⁶⁰ˣ	**8**⁸
1¹⁴ˣ	**2**	**7**	**6**	**4**	**9**	**3**	**8**	**5**

39

3-		1-		48×	360×	8-		19+
5	**2**	**7**	**6**	**4**	**8**	**1**	**9**	**3**
4¹⁰⁺	**6**	**3**¹⁶⁺	**8**	**2**	**5**	**9**	**7**	**1**⁶⁻
8⁵⁻	**3**	**5**	**4**¹⁻	**6**	**9**	**2**²÷	**1**	**7**
3¹³⁺	**8**⁸	**4**²⁴ˣ	**5**	**1**¹⁴ˣ	**2**	**7**	**6**²⁰⁺	**9**
9	**1**	**6**	**7**	**8**	**3**⁷⁺	**4**	**2**³²⁰ˣ	**5**
1¹⁴⁺	**7**⁵⁻	**2**	**9**³÷	**3**¹⁷⁺	**6**	**8**	**5**	**4**
7	**9**¹³⁺	**1**⁸⁻	**3**	**5**¹⁵⁺	**4**	**6**	**8**	**2**⁴÷
6	**4**	**9**	**2**⁹⁺	**7**	**1**¹⁵ˣ	**5**	**3**	**8**
2²	**5**¹³⁺	**8**	**1**¹⁰⁺	**9**	**7**⁴⁻	**3**	**4**	**6**

40

120×	8×	2-		162×		7+		2-
5	**8**	**2**	**4**	**3**	**9**	**1**	**6**	**7**
3	**1**	**7**²⁻	**9**	**6**	**2**¹¹²ˣ	**8**	**4**¹⁸⁰ˣ	**5**
8	**9**⁸⁻	**1**	**2**²⁰⁺	**4**	**6**	**7**	**5**	**3**²÷
1¹⁴ˣ	**2**	**3**⁵⁻	**8**	**7**³¹⁵ˣ	**5**¹⁻	**4**	**9**	**6**
2⁵⁴ˣ	**7**	**8**	**5**	**9**	**1**¹²ˣ	**6**	**3**¹⁻	**4**
9	**3**²¹⁰ˣ	**5**	**6**	**1**⁴÷	**4**	**2**	**7**¹⁷⁺	**8**
4²⁻	**6**	**9**⁹	**7**	**5**¹⁶⁺	**8**	**3**	**2**	**1**¹³⁺
7¹⁸⁺	**4**⁴	**6**²⁴ˣ	**1**	**2**¹³⁺	**3**	**5**¹⁸⁹ˣ	**8**¹³⁺	**9**
6	**5**	**4**	**3**	**8**	**7**	**9**	**1**	**2**

41

2÷ 1	2	240× 5	84× 3	7	13+ 8	3− 9	6	21+ 4
800× 5	6	8	4	3	2	7	9	1
8	5	24× 4	6	19+ 9	19+ 3	1	1− 2	1− 7
4	15+ 7	3÷ 3	9	2	6	1− 5	1	8
3	4	2÷ 1	3− 5	8	9	6	42× 7	2
504× 9	1	2	210× 8	6	60× 7	4	5	3
7	8	23+ 9	1− 2	1	5	3	24× 4	6
18× 2	9	6	16+ 7	4	8× 1	8 8	17+ 3	5
3− 6	3	7	1	5	4	2	8 8	9

42

36× 4	1	3	7+ 8	7	6	90× 5	9	2
1− 5	9	4	84× 7	3	11+ 2	8	12× 6	1
6	3 3	2− 7	9	4	4− 8	1	2	270× 5
4÷ 8	2	1− 5	6	15+ 1	4	81× 3	28× 7	9
16+ 2	7	6+ 1	5	8	3	9	4	6
7	270× 5	18× 9	15× 3	6	1	10+ 2	24× 8	4− 4
9	6	2	1	5	7	2− 4	3	8
1− 3	4	48× 8	15+ 2	9	90× 5	6	11+ 1	7
7− 1	8	6	4	2	9	2− 7	5	3

43

56× 7	20× 4	1	5	20+ 9	6	13+ 8	3	2
4	2÷ 1	2	2− 8	6	5	108× 3	9	2− 7
1	2	21× 3	7 7	12+ 8	1− 9	2− 6	4	5
1− 5	6	7	1	3	8	4	7− 2	9
168× 6	7	4	11+ 9	2	4÷ 1	360× 5	8	15× 3
2	3	360× 8	15+ 6	7	4	9	5	1
1− 8	5	9	2	3÷ 1	3	5− 7	5− 6	10+ 4
9	216× 8	22+ 5	84× 3	4	7	2	1	6
3	9	6	4	5	2	1 1	7	1− 8

44

8× 1	405× 9	5	2÷ 3	6	8	24+ 7	2	13+ 4
4	2	9	3− 5	8	3× 1	3	6	7
9+ 5	4	336× 7	5− 6	1	3	144× 9	8	11+ 2
3− 9	8	6	1	4− 3	7	2	4	5− 5
6	4− 7	3	18+ 9	5	8× 2	4	1	8
7− 8	1	2÷ 2	4	1344× 7	6	315× 5	9	3
1− 2	3	1	8	4	1− 5	7+ 6	7	162× 9
12+ 7	5	2÷ 8	5− 2	18× 9	4	1	3	6
2÷ 3	6	4	7	2	9	1− 8	4− 5	1

45

8	9	1	5	2	4	3	6	7
3	2	9	7	1	8	5	4	6
5	7	3	8	9	6	2	1	4
9	8	4	1	3	5	6	7	2
4	6	2	3	7	9	8	5	1
2	5	6	4	8	1	7	9	3
1	3	5	2	6	7	4	8	9
6	4	7	9	5	2	1	3	8
7	1	8	6	4	3	9	2	5

46

9	8	5	7	1	3	6	2	4
3	4	2	8	9	1	5	6	7
4	2	1	9	5	8	7	3	6
6	3	4	2	7	5	9	1	8
7	5	6	4	2	9	1	8	3
1	6	7	5	8	2	3	4	9
8	9	3	6	4	7	2	5	1
2	1	9	3	6	4	8	7	5
5	7	8	1	3	6	4	9	2

47

3	2	6	4	5	1	8	9	7
5	1	4	8	9	3	7	6	2
1	3	9	6	4	8	2	7	5
8	5	2	3	7	4	9	1	6
2	4	1	7	3	5	6	8	9
4	8	7	9	1	6	5	2	3
6	7	8	5	2	9	4	3	1
7	9	5	1	6	2	3	4	8
9	6	3	2	8	7	1	5	4

48

7	2	9	3	6	1	5	8	4
4	3	5	8	7	6	9	2	1
6	4	3	5	2	9	7	1	8
3	8	4	6	1	7	2	5	9
9	5	1	4	3	8	6	7	2
5	6	7	9	8	2	1	4	3
2	7	8	1	9	4	3	6	5
1	9	6	2	4	5	8	3	7
8	1	2	7	5	3	4	9	6

49

2	5	8	1	3	6	4	7	9
4	1	3	7	8	2	5	9	6
3	2	1	4	5	9	7	6	8
5	6	2	9	4	7	1	8	3
6	8	5	3	2	4	9	1	7
9	4	6	5	7	3	8	2	1
7	3	4	8	9	1	6	5	2
1	9	7	2	6	8	3	4	5
8	7	9	6	1	5	2	3	4

50

7	6	5	3	2	4	1	8	9
3	9	2	6	1	5	4	7	8
2	1	3	8	9	6	5	4	7
4	3	6	5	7	8	9	2	1
1	2	8	9	4	7	6	5	3
9	4	1	7	5	3	8	6	2
5	7	4	2	8	9	3	1	6
8	5	9	1	6	2	7	3	4
6	8	7	4	3	1	2	9	5

51

2	3	4	7	8	1	5	9	6
5	9	3	4	2	7	6	1	8
3	1	9	8	4	6	7	5	2
4	2	1	6	5	9	3	8	7
7	4	2	1	3	8	9	6	5
1	7	6	5	9	4	8	2	3
6	5	8	9	1	3	2	7	4
9	8	7	2	6	5	4	3	1
8	6	5	3	7	2	1	4	9

52

5	3	9	2	8	4	7	1	6
3	2	6	4	1	8	9	7	5
9	6	7	5	3	2	8	4	1
7	1	2	6	9	3	5	8	4
8	5	3	7	6	1	4	9	2
2	7	4	9	5	6	1	3	8
6	9	1	8	4	7	2	5	3
4	8	5	1	2	9	3	6	7
1	4	8	3	7	5	6	2	9

53

13+ 1	9+ 4	4- 3	9	20+ 6	5- 7	2	17+ 8	6+ 5
4	2	7	6	8	8+ 5	3	9	1
8	3	5+ 1	9+ 5	4	13+ 6	7	15+ 2	15+ 9
17+ 5	9	4	2- 3	1	6+ 2	17+ 8	7	6
3	15+ 8	13+ 5	7	2	1	9	6	14+ 4
16+ 9	6	8	10+ 1	7	3	1- 5	4	2
7	1	3- 9	2	18+ 5	12+ 4	7+ 6	2- 3	8
4- 2	12+ 7	6	4	9	8	1	5	10+ 3
6	5	10+ 2	8	6- 3	9	5+ 4	1	7

54

6 6	2- 2	4	17+ 9	18+ 8	6+ 1	5	11+ 7	3
2- 3	18+ 7	2	8	4	6	4- 9	5	1
1	6	8+ 3	5	7 7	19+ 9	8	2	3- 4
7+ 5	3	8- 9	1	6- 2	8	10+ 6	4	7
2	7+ 1	6	3	2- 9	7	4	19+ 8	5
2- 7	5	8	4	6+ 1	2	3	9 9	6
12+ 4	5- 9	9+ 7	2	14+ 6	5	2- 1	3	8 8
8	4	4- 1	13+ 7	12+ 5	3	4- 2	6	7- 9
17+ 9	8	5	6	3	4	8+ 7	1	2

55

3- 3	13+ 4	7	2	7- 8	1	20+ 6	5	9
6	7- 1	16+ 3	7	6- 2	8	4	9	3- 5
8+ 7	8	14+ 9	6	13+ 4	5	4+ 3	1	2
1	1- 2	5	1- 4	9	7	22+ 8	9+ 3	6
14+ 2	3	23+ 6	5	3- 7	4	9	8 8	11+ 1
4	21+ 7	8	9	5- 1	6	5	4- 2	3
8	9	6+ 4	4+ 3	14+ 5	5+ 2	1	6	7
14+ 9	5	2	1	6	23+ 3	7	4	8
5	15+ 6	1	8	3	18+ 9	2	7	4

56

12+ 3	9	2- 6	11+ 8	7 7	9+ 2	5+ 4	1	8+ 5
14+ 6	7+ 4	8	2	1	7	4- 5	9	3
8	2	1	2- 3	5	22+ 6	7	13+ 4	9
11+ 4	10+ 7	3	17+ 5	6	9	3+ 1	2	9+ 8
7	4- 8	4	6	13+ 9	3	4+ 2	13+ 5	1
2- 5	3	7- 2	9 9	4	1	6	15+ 8	7
5- 1	6	9	17+ 7	2	5	11+ 8	3	6+ 4
11+ 9	11+ 5	7	3- 1	8	18+ 4	3	6	2
2	1	5	5- 4	3	8	1- 9	7	6

57

6−	25+	13+	5+	5−	16+	10+		3−
1	8	3	2	9	7	6	4	5
7	6	1	3	4	9	5 (3−)	2 (9+)	8
8	3	9	4 (2−)	5 (1−)	6	2	7	1 (15+)
3 (3)	4 (2−)	2	6	7 (9+)	1 (4−)	8 (4−)	5	9
6 (4−)	1 (4−)	8 (8)	9 (8−)	2	5	4	3 (19+)	7
2	5	4 (2−)	1	6 (6)	8 (15+)	7	9	3 (13+)
9 (16+)	7	6	5 (20+)	3 (1−)	2	1 (9+)	8	4
4 (1−)	2 (3−)	5	7	8	3 (7+)	9 (6−)	1 (7+)	6
5	9 (16+)	7	8 (7−)	1	4	3	6	2 (2)

58

5+		2−	12+	19+		22+		
1	3	5	4	8	2	9	7	6
6 (15+)	1	3	8	2	7	5 (14+)	9 (5−)	4
9	6 (6)	2 (4−)	5 (3−)	3 (12+)	1 (2−)	8	4 (4−)	7 (2−)
7 (11+)	4	6	2	9	3	1	8	5
4 (12+)	9 (16+)	1 (11+)	7 (7)	5 (1−)	6 (2−)	2 (5+)	3	8 (5−)
5	7	9	1	6	8	4 (3−)	2 (8+)	3
3	8 (12+)	4	9 (10+)	1	5 (5)	7	6	2 (8+)
8 (6−)	2 (5−)	7	3 (1−)	4	9 (16+)	6 (6)	5	1
2	5 (3−)	8	6 (1−)	7	4	3 (8−)	1	9

59

11+		15+		5+	16+			17+
7	4	9	6	2	5	1	3	8
6 (6)	8 (17+)	1	2 (1−)	3	7	5 (9+)	4	9
2 (17+)	7	8	3 (5)	5 (15+)	6	9	1 (3−)	4
5	3	2 (1−)	1	4 (13+)	9	6 (9+)	8 (15+)	7
8 (17+)	9 (9+)	5	4 (4)	7 (5−)	2	3	6 (5−)	1
3 (1−)	6 (11+)	4	5 (22+)	8	1 (7−)	7 (16+)	9	2 (1−)
4	5	6 (16+)	9	1 (5−)	8	2 (9+)	7	3
9 (10+)	1 (1)	3	7	6 (16+)	4	8 (7+)	2	5 (1−)
1	2 (17+)	7	8	9 (12+)	3	4	5	6

60

24+			7+			6−		16+
7	9	8	2	4	1	3	5	6
8 (23+)	4 (4−)	7 (6−)	3 (8+)	6 (4−)	2	9	1 (8+)	5
6	8	1	5	9 (10+)	7 (5−)	2	4	3
9	7 (4−)	6 (19+)	8	1	5 (12+)	4	3	2 (3+)
4 (16+)	3	5	9 (30+)	8	6	7	2 (9+)	1
2	6	4	1 (10+)	5 (5)	3 (5−)	8 (7−)	7	9 (13+)
3 (2−)	5 (14+)	9	7	2	8	1	6 (3−)	4
1	2 (5+)	3	6 (13+)	7	4 (16+)	5 (11+)	9	8 (23+)
5 (4−)	1	2 (6+)	4	3	9	6	8	7

61

(16+) 6	9	1	(6−) 8	2	3	(10+) 7	(11+) 4	(16+) 5
(8) 8	(8+) 1	(14+) 2	3	9	7	(14+) 5	(15+) 6	4
(8+) 3	4	(2−) 5	(7−) 1	8	6	2	9	7
5	3	7	(8+) 2	6	1	(4−) 4	8	(14+) 9
(20+) 7	5	8	(3−) 4	1	9	(15+) 6	(3+) 2	3
(10+) 4	6	(14+) 9	5	(17+) 7	8	(9+) 3	1	2
(8−) 9	(5−) 8	3	(3−) 7	4	2	1	5	(14+) 6
1	(11+) 2	(13+) 4	6	3	(17+) 5	(16+) 9	7	8
2	7	(6) 6	(14+) 9	5	4	8	(4+) 3	1

62

(6) 6	(4−) 4	8	(5+) 3	2	7	(16+) 5	(4−) 9	(6+) 1
(10+) 9	(13+) 6	7	(6−) 2	4	5	(9+) 8	1	3
1	7	(9+) 6	(9+) 8	(5) 5	(5−) 9	4	(8+) 3	2
(1−) 4	2	3	(6) 6	(11+) 8	(8−) 1	9	5	(1−) 7
5	9	(8−) 1	4	3	(2−) 8	(18+) 7	(2) 2	6
(12+) 2	1	4	(18+) 5	7	6	3	8	(5−) 9
3	(22+) 8	5	9	6	(5+) 2	(8+) 1	7	4
7	5	2	(8−) 1	9	3	(4−) 6	(2−) 4	(3−) 8
(5−) 8	3	(17+) 9	7	1	(4) 4	2	6	5

63

(3−) 5	(2−) 7	(5−) 6	1	(1−) 8	(4+) 3	(5−) 9	(7+) 2	(4) 4
8	9	(11+) 5	6	7	1	4	3	2
(14+) 1	(15+) 8	(5+) 2	3	(14+) 9	(2−) 4	6	(12+) 7	5
9	6	(4−) 4	2	3	(17+) 8	5	1	7
4	1	8	(11+) 5	6	2	(7) 7	(6−) 9	3
(8+) 3	5	(22+) 9	4	(1−) 1	7	(22+) 2	6	8
(16+) 7	3	(11+) 1	9	2	(20+) 5	(9+) 8	(4) 4	6
2	4	3	(12+) 7	5	6	1	(18+) 8	9
(8+) 6	2	7	(4−) 8	4	9	(2−) 3	5	1

64

(14+) 9	5	(8+) 1	(6−) 8	7	(1−) 2	3	(11+) 6	4
(6−) 8	(15+) 9	7	2	6	(12+) 4	5	3	1
2	6	(5) 5	(1−) 3	4	(23+) 9	8	(6−) 1	(9+) 7
(13+) 4	1	(14+) 3	9	(20+) 5	8	6	7	2
(6) 6	8	2	(2−) 1	3	7	(18+) 4	9	5
(15+) 7	(3) 3	(9+) 4	5	(17+) 9	1	(4+) 2	(2−) 8	6
3	(14+) 4	(22+) 9	7	2	6	1	(9+) 5	(1−) 8
5	2	8	6	(7−) 1	(4−) 3	7	4	9
(8+) 1	7	(2−) 6	4	8	(14+) 5	9	(1−) 2	3

65

1−4	3	12+7	23+6	12+9	6+1	5	2	6−8
11+9	2	5	8	3	18+6	3−7	4	1
9+2	7	8	1	17+5	3	9	10+6	4
12+7	5	7−2	9	8	4	3−3	2−1	4−6
9+5	4	2−9	7	9+1	8	6	3	2
10+1	9	15+4	5	6	7	5−2	18+8	3
3−3	6	4+1	4 4	3−2	5	1−8	9	7
14+6	8	3	12+2	4 4	7−9	3−1	12+7	5
8 8	7+1	6	3	7	2	4	14+5	9

66

8+5	3	2−2	15+7	8	15+6	9	11+1	14+4
15+9	10+1	4	1−2	3	11+5	6	8	7
6	9	2−5	1−4	7 7	7−8	1	2	3
6−1	14+6	7	5	11+4	2	3	1−9	8
7	8	2−1	3	2	7+4	5	6	8−9
4−4	3−2	14+6	24+8	9	3	16+7	5	1
8	5	6−3	8+1	6	9	4	5−7	2
6+2	4	9	6	1	7 7	14+8	14+3	5
4−3	7	1−8	9	4−5	1	2	4	6

67

23+7	8	1−2	1	12+3	9 9	15+6	5	5−4
8	15+4	5	6	2	7	1	3	9
13+9	4+3	1−7	8	5−1	13+5	9+4	17+6	2
4	1	6+3	2−2	6	8	5	9	12+7
8+6	2	1	4	13+9	3	23+7	8	5
2	24+9	6	12+7	5	1	8	7+4	3
2−5	7	9	3 3	12+8	4	12+2	1	6
9+1	5	18+4	9	11+7	4−6	3	6−2	8
3	14+6	8	5	4	2	9 9	6−7	1

68

23+5	4	16+8	1	6−3	9	5−6	13+7	2
8	6	7	9 9	8+2	5	1	16+3	4
5−9	1−3	4	10+2	8	1	7	6	18+5
4	12+1	6	12+3	5	2 2	17+8	9	7
22+7	8	5	4	1 1	7+3	7−9	2	6
4−6	7	5+2	3−8	15+9	4	6+5	1	4+3
2	7−9	3	5	6	23+7	12+4	8	1
2−1	2	16+9	7	3−4	6	16+3	5	8
3	12+5	1	6	7	8	2	5−4	9

69

1-		17+	17+		3-	7-		16+
1	**2**	**9**	**3**	**6**	**7**	**8**	**4**	**5**
16+						2-		
9	**7**	**8**	**6**	**2**	**4**	**1**	**5**	**3**
14+	8-	2-	7-		15+	2		
5	**9**	**3**	**1**	**8**	**6**	**2**	**7**	**4**
			11+			12+	2-	
2	**1**	**5**	**7**	**4**	**9**	**3**	**6**	**8**
	10+			1-	3+		3-	
7	**8**	**2**	**9**	**5**	**1**	**4**	**3**	**6**
21+	3-						10+	
4	**3**	**6**	**8**	**7**	**2**	**5**	**1**	**9**
	10+		3-	7-		29+	10+	
3	**6**	**4**	**5**	**1**	**8**	**9**	**2**	**7**
	3-			3				
8	**4**	**7**	**2**	**3**	**5**	**6**	**9**	**1**
	4-		13+		4-		6-	
6	**5**	**1**	**4**	**9**	**3**	**7**	**8**	**2**

70

14+		2-	19+	20+			6+	
2	**4**	**7**	**6**	**9**	**3**	**8**	**5**	**1**
				15+		5+		2-
3	**5**	**9**	**2**	**7**	**8**	**1**	**4**	**6**
11+	5-			5+	12+		16+	
4	**1**	**2**	**9**	**3**	**5**	**7**	**6**	**8**
		7+			11+	4-		9+
7	**6**	**3**	**1**	**2**	**4**	**9**	**8**	**5**
18+	4-	3-						
9	**7**	**8**	**3**	**1**	**6**	**5**	**2**	**4**
			18+		9+		8-	7-
8	**3**	**5**	**7**	**6**	**2**	**4**	**1**	**9**
	1-	10+	17+		7			
1	**8**	**6**	**4**	**5**	**7**	**3**	**9**	**2**
11+					3+		3	17+
6	**9**	**4**	**5**	**8**	**1**	**2**	**3**	**7**
	3+		4-		15+			
5	**2**	**1**	**8**	**4**	**9**	**6**	**7**	**3**

71

7-	3-	3+	18+				18+	
9	**1**	**2**	**3**	**6**	**4**	**5**	**8**	**7**
				5	26+			2-
2	**4**	**1**	**7**	**5**	**9**	**8**	**3**	**6**
9+	3-	4-		1-	2-		11+	
1	**6**	**7**	**2**	**3**	**5**	**9**	**4**	**8**
			6+			6		
8	**9**	**3**	**1**	**4**	**7**	**6**	**2**	**5**
17+				1-		6+		
4	**7**	**6**	**5**	**9**	**8**	**2**	**1**	**3**
16+			9	9+	11+		2-	
3	**8**	**5**	**9**	**2**	**1**	**7**	**6**	**4**
2-		16+					14+	2
7	**5**	**8**	**4**	**1**	**6**	**3**	**9**	**2**
3-			23+	5-		5+		17+
6	**3**	**4**	**8**	**7**	**2**	**1**	**5**	**9**
3-				5-				
5	**2**	**9**	**6**	**8**	**3**	**4**	**7**	**1**

72

4-	1-	3-	17+		5-		5-	
5	**4**	**3**	**8**	**9**	**2**	**7**	**1**	**6**
			10+	5-	1-		4-	
1	**5**	**6**	**4**	**2**	**8**	**9**	**7**	**3**
17+	7-			4-			7+	5-
8	**9**	**2**	**6**	**7**	**5**	**1**	**3**	**4**
		8+		4+	3-	11+		
2	**7**	**1**	**5**	**3**	**6**	**8**	**4**	**9**
2-	13+						3-	
4	**6**	**7**	**2**	**1**	**9**	**3**	**5**	**8**
	20+			9+			8-	
6	**8**	**5**	**7**	**4**	**3**	**2**	**9**	**1**
16+	2-	12+	20+		11+	3-		2
7	**3**	**4**	**9**	**6**	**1**	**5**	**8**	**2**
			2-			4-		2-
9	**1**	**8**	**3**	**5**	**4**	**6**	**2**	**7**
1-		9	19+					
3	**2**	**9**	**1**	**8**	**7**	**4**	**6**	**5**

73

6+ 4	2	13+ 6	7	20+ 3	8	3− 9	1	16+ 5
8− 1	1− 7	8	7− 9	2	5	4− 6	4	3
9	8+ 5	3	2	7	1	21+ 4	6 6	8
2− 5	3	2− 2	4	4− 9	6	7	1− 8	3+ 1
2− 6	8	14+ 9	13+ 3	5	4	8+ 1	7	2
7− 8	1	5	6	4	3 3	2	2− 9	7
5− 7	5+ 4	1	14+ 8	6	19+ 2	5	12+ 3	9
2	15+ 6	6− 7	1	8	9	3 3	15+ 5	4
3 3	9	17+ 4	5	1	7	10+ 8	2	6

74

4− 1	5	5− 3	8	3− 4	22+ 6	9	3− 7	5+ 2
3− 2	6	30+ 8	9	1	13+ 5	7	4	3
5	2	9	4	6	8	3− 1	4+ 3	2− 7
15+ 8	7	21+ 6	9+ 2	9	6+ 1	3	5	4
4 4	9	5	7	5− 8	3	2	5− 1	6
2− 9	1	4	6	3	7	12+ 5	8+ 2	17+ 8
7	6+ 3	2	1	12+ 5	13+ 4	8 8	6	9
1− 3	4	8+ 7	5	2	9	12+ 6	9+ 8	1
2− 6	8	1	4− 3	7	2	4	14+ 9	5

75

6+ 4	2	14+ 5	3	5− 8	7	2− 9	3− 6	12+ 1
7− 1	8	2− 2	6	3	9	10+ 5	7	4
1− 6	16+ 3	4	14+ 1	5	8	2	19+ 9	13+ 7
5	4	9	14+ 7	3+ 1	2	3	8	6
6− 7	1	7+ 6	5	13+ 4	3	13+ 8	2	17+ 9
26+ 8	9	1	2	7	6	4	3	5
9	10+ 7	3	21+ 8	6	10+ 5	1	3− 4	6− 2
11+ 3	20+ 5	7	9	7− 2	4	6 6	1	8
2	6	8	4	9	1	2− 7	5	3 3

76

20+ 2	9	4	9+ 6	3	8+ 5	16+ 7	16+ 8	1
5	13+ 4	8	3	6+ 2	9+ 6	9	18+ 1	7
15+ 7	7+ 5	1	8 8	4	3	2	6	9
8	2	18+ 9	4	5	8+ 7	1	11+ 3	6
13+ 4	4+ 3	6 6	14+ 9	8+ 7	1	5	2	11+ 8
9	1	9+ 2	5	12+ 8	4	6	7 7	3
10+ 3	20+ 8	7	7+ 1	6	13+ 9	4	7+ 5	2
6	7	5	9+ 2	1	8	12+ 3	9	9+ 4
1	9+ 6	3	7	11+ 9	2	12+ 8	4	5

77

3 ⁱ³⁺	2	8	4 ¹⁰⁺	1 ⁴⁺	5 ¹⁸⁺	7	6	9 ¹⁷⁺
4 ²⁰⁺	8	2 ²	6	3	1 ⁵⁺	5 ¹¹⁺	9 ¹⁴⁺	7
8	9 ¹⁸⁺	3	7 ⁹⁺	2	4	6	5	1
9 ¹⁶⁺	6	1 ⁴⁺	2	7 ¹⁹⁺	8 ¹⁵⁺	4	3 ⁵⁺	5 ¹⁵⁺
5	7 ¹⁰⁺	6 ²⁰⁺	1	8	9 ²⁴⁺	3	2	4
2	3	5	9	4	7	8	1 ¹⁰⁺	6
7 ¹⁴⁺	4 ⁵⁺	9 ¹⁶⁺	3 ⁸⁺	5 ¹¹⁺	6	1	8	2 ¹⁶⁺
6	1	7	5	9 ¹⁵⁺	3 ⁵⁺	2	4	8
1	5 ¹⁷⁺	4	8	6	2	9 ¹⁹⁺	7	3

78

9 ¹³⁺	1 ¹	6 ⁹⁺	4 ⁷⁺	5 ¹²⁺	3 ⁹⁺	7 ¹⁸⁺	8 ¹⁰⁺	2
4	2	1	3	7	6	8	5 ⁹⁺	9 ¹⁶⁺
8 ¹⁸⁺	9 ¹⁴⁺	5	6 ⁶	2 ⁶⁺	1 ¹¹⁺	3	4	7
3	6 ¹⁷⁺	7	9 ¹⁴⁺	4	8	2	1 ¹⁴⁺	5
7	4	9 ¹⁷⁺	5	3 ⁵⁺	2	1 ⁷⁺	6	8
1 ⁴⁺	3	8	7 ¹³⁺	6	5 ²³⁺	9	2	4 ⁵⁺
6 ¹³⁺	8 ¹⁵⁺	3 ⁵⁺	2	9 ¹³⁺	4	5 ²¹⁺	7	1
5	7	2 ²	1 ¹⁷⁺	8	9	4	3	6 ⁹⁺
2	5 ⁹⁺	4	8	1 ¹⁴⁺	7	6	9 ⁹	3

79

8 ²⁷⁺	7 ⁹⁺	2	4	3 ⁷⁺	5 ²⁵⁺	9	6	1 ⁷⁺
4	8	7 ¹⁰⁺	3	9 ¹⁴⁺	6 ¹⁷⁺	5	1 ⁴⁺	2
7	2 ¹⁰⁺	8	6 ¹³⁺	5	9	1 ⁷⁺	3	4
1 ⁷⁺	3 ⁸⁺	5	7	8 ¹³⁺	2	6	4 ¹²⁺	9 ¹⁶⁺
6	1 ⁵⁺	9 ¹⁴⁺	5	2	3	4 ¹²⁺	8	7
5 ²⁰⁺	4	3 ⁴⁺	9 ¹⁷⁺	7 ¹²⁺	1	8	2 ⁹⁺	6 ⁶
9	6	1	8	4	7 ²¹⁺	2	5	3 ⁸⁺
2 ¹¹⁺	9	4 ⁵⁺	1	6	8	3 ¹⁰⁺	7 ²⁴⁺	5
3 ⁸⁺	5	6 ⁸⁺	2	1 ⁵⁺	4	7	9	8

80

5 ⁹⁺	2 ⁹⁺	7	6 ¹⁴⁺	8 ¹³⁺	4 ⁵⁺	1	9 ¹²⁺	3
4	9 ¹²⁺	3	8	5	2 ¹¹⁺	7	1 ⁷⁺	6
3 ⁹⁺	6	1 ¹⁸⁺	5 ⁷⁺	4 ¹²⁺	8	2	7 ¹⁶⁺	9
6 ⁷⁺	7 ¹¹⁺	9	2	3 ¹⁶⁺	5	8	4 ⁵⁺	1
1	4	8 ¹⁰⁺	3	7	9 ²⁴⁺	6	5 ⁷⁺	2
7 ¹⁵⁺	8	5 ⁶⁺	1	6	3	9 ⁹	2 ⁶⁺	4
8 ¹⁹⁺	5 ⁹⁺	4 ¹³⁺	9	2 ³⁺	1 ⁴⁺	3	6	7 ¹²⁺
9	3	2 ⁸⁺	7 ²⁰⁺	1	6 ²³⁺	4	8	5
2	1	6	4	9	7 ⁷	5	3 ¹¹⁺	8

81

4	9	8	2	7	3	1	5	6
3	5	9	4	1	2	7	6	8
1	7	6	9	8	4	2	3	5
5	4	1	3	2	6	8	7	9
9	6	4	8	3	7	5	1	2
7	2	5	6	9	1	4	8	3
8	1	7	5	6	9	3	2	4
2	8	3	7	4	5	6	9	1
6	3	2	1	5	8	9	4	7

82

9	4	2	5	6	3	7	1	8
3	5	7	2	1	4	9	8	6
5	8	6	7	2	1	4	9	3
2	9	4	1	7	8	6	3	5
4	3	5	9	8	2	1	6	7
1	2	3	4	5	6	8	7	9
6	1	9	8	4	7	3	5	2
8	7	1	6	3	9	5	2	4
7	6	8	3	9	5	2	4	1

83

3	2	9	4	6	5	7	1	8
4	6	8	9	3	7	2	5	1
1	5	6	3	4	2	8	9	7
5	7	1	2	8	6	3	4	9
9	4	3	7	2	1	6	8	5
6	9	2	5	1	8	4	7	3
7	8	5	6	9	4	1	3	2
2	1	7	8	5	3	9	6	4
8	3	4	1	7	9	5	2	6

84

9	7	3	4	8	1	6	5	2
4	5	8	3	9	7	2	1	6
6	2	5	8	3	9	1	4	7
1	3	4	5	7	6	8	2	9
7	6	2	9	4	8	5	3	1
8	4	9	1	5	2	7	6	3
5	1	6	7	2	4	3	9	8
3	8	1	2	6	5	9	7	4
2	9	7	6	1	3	4	8	5

85

9+		17+	16+		8+	10+	12+	
2	3	6	7	9	5	1	4	8
4	6	5	7+ 2	1	3	9	18+ 8	7
7+ 1	2	25+ 9	4	14+ 8	6	12+ 5	7	3
17+ 5	4	7	9	16+ 3	10+ 8	2	11+ 6	7+ 1
9	6+ 1	12+ 4	8	6	10+ 7	3	5	2
3	5	7+ 1	6	7	24+ 9	8	11+ 2	4
23+ 8	11+ 9	2	6+ 1	5	4	10+ 7	3	6
7	8	11+ 3	8+ 5	4	2	11+ 6	1	23+ 9
13+ 6	7	8	3+ 3	2	1	4	9	5

86

6+	13+		12+	20+		16+	11+	9+
1	9	2	4	8	3	7	5	6
5	2	7	1	13+ 4	9	8	6	3
8+ 6	7+ 3	4	8	17+ 9	7	12+ 1	2	7+ 5
2	7+ 1	16+ 8	3	6	5	11+ 4	7	13+ 9
16+ 9	6	3	7 5	7	13+ 8	2	5+ 1	4
7	5	11+ 6	9	15+ 1	4+ 2	3	4	15+ 8
3	4	15+ 9	6	5+ 2	1	13+ 5	8	7
12+ 8	15+ 7	5	14+ 2	3	10+ 4	27+ 6	9	3+ 1
4	8	1	7	5	6	9	3	2

87

12+	11+			12+		13+	24+	2
3	6	1	4	5	7	8	9	2
4	6+ 2	19+ 9	10+ 6	1	3	5	8	7
5	4	3	7	6+ 2	10+ 8	15+ 9	13+ 6	9+ 1
17+ 9	6+ 1	9+ 5	12+ 3	4	2	6	7	8
8	5	4	9	3 3	5+ 1	10+ 7	7+ 2	15+ 6
7+ 1	15+ 8	14+ 6	3+ 2	16+ 7	4	3	5	9
6	7	8	1	9	8+ 5	2	7+ 3	12+ 4
14+ 2	3	9+ 7	23+ 8	6	9	1	4	5
7 7	9	2	13+ 5	8	11+ 6	4	1	3

88

6+	22+	10+	15+		9+	13+	11+	
1	8	9	2	6	3	5	4	7
5	2	1	3 3	7	6	8	26+ 9	4
9	3	17+ 8	4	5	3+ 2	1	7	6
17+ 7	9+ 4	8+ 3	1	26+ 9	10+ 8	2	9+ 6	7+ 5
6	5	4	9	8	8+ 1	7	3	2
4	21+ 9	6	12+ 5	10+ 2	7	12+ 3	8	1
8 8	6	14+ 5	7	1	14+ 9	9+ 4	2	3
5+ 3	7	2	10+ 6	4	5	10+ 9	1	8 8
2	1	8+ 7	8	11+ 3	4	10+ 6	14+ 5	9

89

ⁱ⁵⁺6	9	¹⁰⁺2	8	¹⁰⁺3	7	⁶⁺5	1	¹³⁺4
¹⁷⁺4	5	¹²⁺1	¹¹⁺3	8	2	¹⁵⁺7	6	9
⁷7	8	6	¹⁵⁺9	2	¹²⁺5	4	3	⁸⁺1
¹²⁺8	3	5	4	¹⁹⁺6	¹⁷⁺9	⁸⁺1	2	7
1	²⁵⁺4	7	¹¹⁺6	9	8	³3	5	¹²⁺2
¹⁴⁺9	6	8	5	4	⁵⁺1	¹⁶⁺2	7	3
5	¹⁸⁺2	9	7	1	3	8	²¹⁺4	⁶6
⁵⁺2	¹¹⁺7	3	1	⁹⁺5	4	6	9	8
3	⁵⁺1	4	⁹⁺2	7	¹⁵⁺6	9	¹³⁺8	5

90

¹²⁺8	¹⁴⁺4	²⁰⁺9	6	5	⁸⁺7	1	⁵⁺2	3
4	1	⁹⁺7	¹⁰⁺3	¹⁹⁺6	¹⁶⁺5	¹⁹⁺9	8	2
⁷⁺6	9	2	7	4	8	¹⁸⁺5	⁴⁺3	1
1	⁷⁺2	5	⁶⁺4	9	3	7	6	¹⁵⁺8
¹²⁺9	¹³⁺5	8	2	⁹⁺3	6	⁵⁺4	1	7
3	²⁰⁺7	¹⁰⁺6	²²⁺9	³⁺1	2	8	¹³⁺5	¹⁸⁺4
7	6	4	8	²2	⁵⁺1	¹¹⁺3	9	5
⁷⁺2	⁴⁺3	1	5	²⁵⁺8	4	6	¹¹⁺7	¹⁵⁺9
5	¹¹⁺8	3	1	7	9	2	4	6

91

¹³⁺7	6	¹⁸⁺4	8	¹⁴⁺9	5	¹⁸⁺2	¹³⁺1	3
⁵⁺4	¹²⁺3	2	6	⁵5	7	1	8	9
1	7	²¹⁺3	5	4	¹⁸⁺9	⁹⁺6	⁹⁺2	¹⁰⁺8
⁹⁺5	4	¹³⁺6	9	1	8	3	7	2
⁸⁺6	2	7	⁶⁺1	¹²⁺8	4	¹²⁺9	3	⁹⁺5
¹⁷⁺9	¹³⁺5	8	2	⁴⁺3	1	¹⁷⁺7	6	4
8	⁹⁺1	¹⁴⁺5	3	⁸⁺2	6	4	¹⁶⁺9	7
⁵⁺2	8	9	¹³⁺7	6	⁸⁺3	5	⁵⁺4	1
3	¹⁰⁺9	1	²¹⁺4	7	2	8	¹¹⁺5	6

92

⁶⁺1	5	¹⁵⁺7	8	⁸⁺2	6	¹⁷⁺3	9	¹⁸⁺4
¹³⁺4	1	¹⁹⁺9	7	3	¹²⁺2	5	8	6
8	⁵⁺4	1	¹⁶⁺9	7	3	¹¹⁺6	5	⁵⁺2
¹⁶⁺9	7	4	3	²³⁺6	¹³⁺5	8	2	1
¹¹⁺2	6	3	¹¹⁺4	9	8	⁸⁺7	1	¹⁴⁺5
⁸⁺5	⁵⁺3	¹⁹⁺8	6	⁷⁺1	4	2	¹⁹⁺7	9
3	2	6	1	¹³⁺5	⁸⁺7	¹³⁺9	4	8
¹³⁺7	¹⁹⁺9	5	⁷⁺2	8	1	4	⁹⁺6	3
6	8	2	5	¹³⁺4	9	¹1	¹⁰⁺3	7

93

18+ 7	3	8	**14+** 9	5	**8+** 6	**5+** 4	1	**2** 2
10+ 4	5	**4+** 1	3	**18+** 8	2	**15+** 7	6	**22+** 9
1	**9** 9	**14+** 6	8	3	4	2	**14+** 5	7
7+ 5	2	**22+** 4	**13+** 7	6	3	**11+** 8	9	1
12+ 3	8	7	**19+** 6	4	9	1	2	5
9	**10+** 6	3	**14+** 2	**8+** 7	1	**5** 5	**19+** 8	4
16+ 2	4	**15+** 5	1	9	**14+** 8	6	7	**13+** 3
8	1	9	**9+** 5	2	7	**10+** 3	4	6
6	**9+** 7	2	**6+** 4	1	5	**20+** 9	3	8

94

8+ 6	2	**13+** 4	9	**18+** 7	1	**4+** 3	**20+** 8	5
7+ 3	**9+** 1	**13+** 9	6	2	**25+** 4	**6+** 5	7	**14+** 8
4	8	3	2	9	7	1	**16+** 5	6
14+ 9	5	1	**29+** 3	6	8	7	4	**6+** 2
3+ 1	**15+** 9	6	7	8	5	**5+** 2	3	4
2	**3** 3	**14+** 7	8	5	6	**18+** 4	**10+** 9	1
12+ 5	4	**10+** 8	1	**8+** 3	**12+** 9	6	**3+** 2	**16+** 7
7	**21+** 6	2	5	4	3	8	1	9
8	7	**9+** 5	4	1	**11+** 2	9	**9+** 6	3

95

7+ 1	2	4	**8+** 3	5	**15+** 7	8	**16+** 6	**16+** 9
15+ 5	1	**18+** 2	8	**12+** 9	3	6	4	7
9	**24+** 8	5	2	6	**11+** 4	7	**5+** 1	3
6+ 4	**12+** 3	6	5	**17+** 2	**8** 8	**16+** 9	7	1
2	9	**7+** 3	4	7	**6+** 1	5	**11+** 8	**8+** 6
20+ 6	**6+** 5	1	**7** 7	8	**21+** 9	4	3	2
8	6	**16+** 7	9	**11+** 1	2	3	5	**18+** 4
10+ 7	**19+** 4	8	**4+** 1	3	6	2	9	5
3	7	**15+** 9	6	**9+** 4	5	**11+** 1	2	8

96

8+ 1	7	**11+** 3	8	**9+** 5	4	**8+** 6	2	**12+** 9
20+ 5	**3+** 2	1	**7+** 4	**10+** 8	**7+** 6	**16+** 7	9	3
6	9	**13+** 7	3	2	1	**15+** 8	4	**20+** 5
9+ 4	5	6	**3+** 2	**18+** 9	8	1	3	7
24+ 9	**8+** 6	2	1	**10+** 3	7	**9+** 4	5	8
7	8	**20+** 9	5	6	**3** 3	**7+** 2	1	4
10+ 8	**4+** 1	**19+** 4	9	**11+** 7	**8+** 5	3	**13+** 6	**3+** 2
2	3	**17+** 8	6	4	**14+** 9	5	7	1
3 3	4	5	**8+** 7	1	**25+** 2	9	8	6

11+ 6	14+ 9	5	15+ 8	7	6+ 1	3+ 2	4	15+ 3
3	2	14+ 9	16+ 7	12+ 4	5	1	8	11+ 6
21+ 8	4	1	9	2	6+ 6	11+ 7	3	5
9	12+ 7	14+ 8	5	6	3	7+ 4	1	2 2
4	5	13+ 6	1	17+ 8	9	3	7+ 2	8+ 7
3+ 2	11+ 3	7	10+ 6	20+ 9	4	8	13+ 5	1
1	6	2	4	16+ 3	7	5	15+ 9	28+ 8
12+ 7	9+ 1	8+ 3	2	5	8	9 9	6	4
5	8	4 4	3	9+ 1	2	6	7	9

9+ 5	7+ 6	6+ 4	10+ 9	1	15+ 7	8	5+ 2	3
4	1	2	8+ 3	5	17+ 9	13+ 7	6	18+ 8
9 9	15+ 3	5	7	18+ 4	8	7+ 2	17+ 1	6
8+ 1	23+ 7	7+ 6	14+ 8	9	2	5	3	4
7	2	1	6	3	7+ 4	13+ 9	8	6+ 5
11+ 6	9	17+ 8	17+ 2	7	3	4	5	1
2	5	5+ 9	1	8	20+ 6	3	4	7
3	15+ 8	7	4	11+ 6	5	7+ 1	16+ 9	11+ 2
12+ 8	4	8+ 3	5	3+ 2	1	6	7	9

8+ 5	18+ 7	9	7+ 6	10+ 2	8	22+ 3	4	6+ 1
3	6+ 4	2	1	15+ 9	6	8	7	5
6+ 4	2	14+ 6	4+ 3	1	16+ 9	8+ 7	18+ 5	8 8
2	5	3	10+ 8	14+ 6	7	1	9	4
1 1	7+ 3	4	2	8	9+ 5	11+ 9	16+ 6	7
21+ 7	7+ 6	9+ 1	14+ 9	5	4	2	8 8	3
6	1	8	12+ 7	16+ 4	3	5	11+ 2	9
8	26+ 9	12+ 7	5	14+ 3	3+ 2	4	4+ 1	8+ 6
9	8	5	4	7	1	6 6	3	2

11+ 6	5	20+ 9	2	9+ 7	9+ 8	9+ 4	4+ 3	1
10+ 3	7	10+ 8	9	2	1	5	10+ 6	4
14+ 5	15+ 8	2	13+ 7	6	14+ 4	18+ 9	1	5+ 3
9	4	10+ 6	6+ 1	5	7	3	8	2
14+ 2	3	4	18+ 8	1	9	10+ 7	5 5	21+ 6
4	11+ 2	3	15+ 6	9	22+ 5	1	7	8
8	6	1 1	9+ 5	4	3	2	16+ 9	7
11+ 1	9	11+ 7	4	11+ 3	6	8	7+ 2	5
7 7	1	8+ 5	3	8	8+ 2	6	13+ 4	9

101

7	3	4	2	8	5	6	9	1
3	7	9	1	6	8	4	2	5
6	9	1	7	3	4	5	8	2
4	2	7	3	1	9	8	5	6
1	6	8	9	5	7	2	3	4
9	1	2	5	4	6	3	7	8
8	5	3	6	7	2	1	4	9
5	8	6	4	2	3	9	1	7
2	4	5	8	9	1	7	6	3

102

4	2	6	3	5	8	7	9	1
3	9	8	6	7	1	4	2	5
1	8	2	7	4	9	5	3	6
8	7	1	9	2	6	3	5	4
7	6	4	2	9	5	1	8	3
6	5	3	1	8	7	9	4	2
5	4	7	8	1	3	2	6	9
2	1	9	5	3	4	6	7	8
9	3	5	4	6	2	8	1	7

103

3	9	2	8	5	6	7	4	1
4	8	6	1	9	2	3	5	7
1	3	4	2	7	5	9	6	8
6	2	9	4	1	8	5	7	3
5	7	1	3	8	4	6	2	9
8	6	7	9	2	3	4	1	5
7	5	8	6	3	1	2	9	4
9	4	3	5	6	7	1	8	2
2	1	5	7	4	9	8	3	6

104

1	3	6	7	9	2	8	4	5
6	7	3	4	8	5	9	1	2
8	2	5	6	3	9	1	7	4
4	1	7	3	2	6	5	8	9
5	8	2	9	7	4	3	6	1
7	5	4	8	1	3	2	9	6
3	6	9	5	4	1	7	2	8
2	9	8	1	6	7	4	5	3
9	4	1	2	5	8	6	3	7

105

960 6	5	23 7	9	24 4	6 1	2	3	1 8
14 5	8	4	7	3	2	5 1	6	9
8	60 2	6	180 4	5	9	20 3	1	7
1	3 6	5	11 3	8	4 7	9	2 2	4
2 4	9	42 1	6	7	3	23 8	105 5	2 2
2	10 1	9	3 8	6	4	5	7	3
13 9	4	8 3	5	3 2	6	30 7	40 8	1
147 7	3	2	16 1	9	8	6	4 4	5
3	7	8	2	4 1	5	4	19 9	6

www.ingramcontent.com/pod-product-compliance
Lightning Source LLC
Chambersburg PA
CBHW071552040426
42452CB00008B/1149